EUROPEAN INTEG
HOUSING POLICY

The European single market and the European single currency are now a reality. Alongside economic globalisation, individual European housing policies are facing many similar pressures. This book examines how these pressures impact on Europe's diverse housing systems; whether European housing systems converge as a result; and what the consequences are of the European Union's further expansion towards the east.

These issues are addressed in four principal sections. First, the consequences of European integration for different housing markets are examined, with separate chapters charting the progress made towards the integration of housing finance markets and the development of a Europe-wide construction industry. Second, a series of policies are examined, such as the impact the Maastricht Treaty's commitment to fiscal restraint is having on the structuring of housing subsidies. Third, the social context of European integration is examined addressing such issues as income distribution and homelessness. The final section provides descriptive accounts of housing in the Nordic countries and in Central and Eastern Europe.

European Integration and Housing Policy exposes a complex series of debates which arise from individual countries' differing housing needs within the confines of European integration. Each member state has a different starting point for demand and supply of both subsidies and housing provision, with more than one strategy open to them. By exploring these disparities, the authors demonstrate that progress towards European integration in housing systems and policies is not a straightforward issue.

Mark Kleinman is Senior Lecturer in Social Policy at the London School of Economics. **Walter Matznetter** is Assistant Professor at the Department of Geography, University of Vienna, and **Mark Stephens** is Lecturer at the Centre for Housing Research and Urban Studies, University of Glasgow.

Routledge/RICS Issues in Real Estate and Housing Series

RICS ISSUES IN REAL ESTATE AND HOUSING SERIES

Series Editors:

Stephen Brown
Managing Editor, RICS

Gerald Brown
University of Salford

John Henneberry
University of Sheffield

Including both leading research and student textbooks, the RICS book series advances an understanding of contemporary issues and debates relating to real estate and housing in both national and global markets.

EUROPEAN INTEGRATION AND HOUSING POLICY
Edited by Mark Kleinman, Walter Matznetter and Mark Stephens

HOUSING, INDIVIDUALS AND THE STATE
The morality of government intervention
Peter King

EUROPEAN INTEGRATION AND HOUSING POLICY

Edited by Mark Kleinman, Walter Matznetter and Mark Stephens

THE ROYAL
INSTITUTION
OF CHARTERED
SURVEYORS

London and New York

First published 1998
by Routledge
11 New Fetter Lane, London EC4P 4EE

Simultaneously published in the USA and Canada
by Routledge
29 West 35th Street, New York, NY 10001

Typeset in Galliard by RefineCatch Limited, Bungay, Suffolk
Printed and bound in Great Britain by
MPG Books Ltd, Bodmin

British Library Cataloguing in Publication Data
A catalogue record for this book is available from the British Library

Library of Congress Cataloging in Publication Data
European integration and housing policy / edited by Mark Kleinman,
Walter Matznetter and Mark Stephens.
    p.   cm. – (Routledge/RICS issues in real estate and housing
series)
1. Housing policy – European Union countries. 2. Europe –
Economic integration. I. Kleinman, Mark. II. Matznetter, Walter,
1952– . III. Stephens, Mark, 1963– . IV. Series.
      HD7332.A3E94   1998
      363.5′094 – dc21   98–13901

ISBN 0–415–17025–7 (hbk)
ISBN 0–415–17026–5 (pbk)

CONTENTS

CONTENTS

FIGURES

TABLES

CONTRIBUTORS

Michael Ball is a Professor of Urban Economics and Policy at South Bank University.

James Barlow is a Professor of Housing and Construction at the Department of the Built Environment, University of Westminster.

Suzanne Fitzpatrick is a Research Fellow at the ESRC Centre for Housing Research and Urban Studies, University of Glasgow.

Marietta Haffner is a researcher at the OTB Research Institute for Housing, Urban and Mobility Studies, Delft University of Technology.

Michael Harloe is Vice-Chancellor of Salford University.

John Hills is Professor of Social Policy and Director of the ESRC Centre for Analysis of Social Exclusion, London School of Economics.

Valerie Karn is Professor of Housing Studies at the University of Manchester and Co-ordinator of the Manchester Housing Research Group.

Mark Kleinman is a Senior Lecturer in Social Policy at the London School of Economics and Political Science.

Walter Matznetter is an Assistant Professor at the Department of Geography, University of Vienna.

Louise Nyström is Director of the Swedish Urban Environment Council.

Michael Oxley is Avebury Professor of Housing at the Centre for Comparative Housing Research, De Montfort University.

Anne Power is a Reader in Housing and Social Policy at the London School of Economics and Political Science.

Hugo Priemus is Director of the OTB Research Institute for Housing, Urban and Mobility Studies, Delft University of Technology.

Jacqueline Smith is a Research Fellow at the Centre for Comparative Housing Research, De Montfort University.

Mark Stephens is Lecturer in European Housing at the ESRC Centre for Housing Research and Urban Studies, University of Glasgow.

Iván Tosics is Co-Director of the Metropolitan Research Institute, Budapest.

Sirpa Tulla is a Research Officer at the Ministry of Environment, Helsinki.

Christine Whitehead is Reader in Housing Economics at the London School of Economics and Political Science.

This book arose from the work of the European Network for Housing Research Working Group on European Integration and National Housing Policies.

1

INTRODUCTION

From comparative housing research to European housing research

Walter Matznetter and Mark Stephens

Introduction: the context of European integration

Anyone who has followed the progress from the European Council meeting in Maastricht in December 1991, towards the creation of the European single currency, due to begin in 1999, would have been struck by the enormous difficulties that have been encountered. Indeed, on several occasions it looked as if the project would be abandoned. The key problem faced by the architects of monetary union has been the unexpectedly deep and long recession in Europe in the first half of the 1990s, which has been followed by the persistence of high unemployment in many countries. These economic difficulties have been associated with the progress towards the single currency in two ways: first, in the dramatic speculation that engulfed the European Exchange Rate Mechanism (the system of currency management within set bands established in 1979) both in 1992, when the UK, Spain and Italy were forced to withdraw, and in 1993 (when the French franc finally succumbed to the speculators); second, in the austerity programmes adopted by national governments to reduce their recession-bloated deficits to meet one of the requirements of the single currency membership. These provoked opposition in many countries, but the outcry had greatest impact in France where strikes and demonstrations erupted at the end of 1995.

During the ratification process, the first Danish referendum went against monetary union, although a second one reversed the decision; and, contrary to expectations, the French gave monetary union only the most grudging vote of approval in their referendum held in 1992. While the Treaty went through some European Parliaments with virtually no discussion, it was bogged down in the UK Parliament for months on end, the issue having split the ruling Conservative Party down the middle. And, when François Mitterrand's presidency ended in 1995, so did the thirteen-year partnership with Chancellor

Kohl, which had given such strength to the pro-integration Franco-German axis.

But, looking back, perhaps in a decade's time, it is conceivable that in the last two decades of the twentieth century, Europe will appear to have undergone an extraordinarily rapid advance in integration. The European Economic Community (EEC) of old was notorious for possessing a Treaty whose objectives were the stuff of pipe dreams. The system by which decisions had to be agreed by all countries meant that it was difficult to secure agreement on anything. It is worth remembering that it took the original six members of the EEC ten years from its foundation in 1958 even to establish a customs union, i.e. a free trade area with a common external tariff. Expansion was achieved in the 1970s with the UK, Ireland and Denmark joining in 1973. But other moves towards integration, such as the currency 'snake', established after the collapse of the post-war Bretton Woods system of (nearly) fixed exchange rates based on the US dollar, ended in disappointment. The early years of the Exchange Rate Mechanism (ERM) were dogged by persistent realignments. Even the European Parliament, which was first elected in 1979, had few formal powers and made little public impact. Once again, the most noticeable achievement up to the mid-1980s was the continued steady expansion in the Community's membership, now to include the recently democratised countries of Greece (1981), Spain and Portugal (both in 1986).

From the mid-1980s, the process of European integration speeded up. Lord Cockfield, then one of the British European Commissioners, drafted a White Paper that proposed the creation of a European single market, described as 'an area without internal frontiers in which the free movement of goods, persons and capital is ensured'. On one level, this document can be seen as proposing technical changes to remove so-called 'non-tariff' barriers and to make free trade within the European Community (EC) (as the EEC was known after 1987) a reality. The sclerotic decision-making process was refined, so that most decisions in the Council of Ministers were made by 'qualified majority voting', which removed the veto held by individual countries. And, instead of detailed regulations being produced (and agreed) for each and every product or service, the White Paper proposed that trade within the EC should take place on the basis of basic standards established by the Council of Ministers ('minimum harmonisation'). The White Paper led to the major Treaty amendments known as the Single European Act in 1987, and, while there were important exceptions, not least in the free movement of people, the bulk of the legislative programme connected with the single market was passed by the end of 1992, the deadline set for the completion of the single market.

But the importance of the single market goes further than this. The Cockfield White Paper conceived of the European Community in terms of the world economy, making it one of the first major decisions influenced by what is now commonly dubbed 'globalisation'. The White Paper emphasised the need for the EC to become not just a single market place, but a single production base, too. As the world economy became more competitive and open, the existence of separate industries in each member state was seen as being inefficient. Far better for members, or regions within them, to maximise scale and agglomeration economies to produce efficient single European industries. To achieve this, it was essential to free the movement of capital, a decision which severely undermined the ability of member states to run independent monetary policies (or indeed resist speculation against currencies in the ERM). Of course there was an element of coincidence in the removal of exchange controls: it must have been easier for members of the EC to abandon exchange controls when world-wide factors, notably the effect of computer technology on the cost and detectability of the movement of money as well as the creation of financial instruments, made their retention implausible in the medium term.

The single market also had important spin-offs in terms of social and regional policy. The Single European Act (SEA) inserted a commitment to economic and social cohesion. Rather than paying for social welfare programmes, such as social security benefits, European social policy takes the form of a commitment to workplace rights, as manifested in the 'social chapter' of the Maastricht Treaty. More importantly, the SEA led the European Commission to establish a rationale for an expansion in its budget and system of regional aid. The Commission argued that, although the shift towards a single market would increase the EC's general level of prosperity, there would be uneven regional and sectoral impacts. In particular, peripheral regions and old industrial regions might be expected to fare less well under the competitive environment created by the single market. Consequently, the European Structural Funds were reformed and expanded. For our purposes, the most relevant funds are the European Regional Development Fund, which aims to promote infrastructure projects which should enhance productive capacity in disadvantaged regions, and the European Social Fund, which is intended to assist workers to gain access to jobs by improving their skills. In the lead-in to the creation of a single currency a new Cohesion Fund was established to support communications and environmental projects in the four poorest member states. By the end of the century, Greece and Portugal are due to receive the equivalent of some 4 per cent of their national incomes in structural assistance from the European Union.

At the time of writing (in late 1997) it looks as if monetary union will proceed, although it is not entirely clear how many countries will qualify for membership initially. Opposition to the single currency has failed to be articulated coherently in Germany and France. The German Social Democrats have been inclined to tap emotional fears about the surrender of the mighty Deutschmark, rather than articulate concerns about unemployment. In France, first the newly elected President Chirac abandoned his concerns regarding the long-term deflationary impact of the *franc-fort* policy of maintaining the value of the French franc against the Deutschmark and has become a reliable supporter of monetary union as the logical development of this policy. The French Socialists, unexpectedly returned to government in the parliamentary elections in 1997 on a platform that explicitly expressed concerns about the impact of European integration on employment levels in France, were very soon placated by sympathetic noises being made at the European Council.

Even in the UK, neither Labour nor Conservative parties was prepared to rule out founder membership of the single currency in the 1997 election. Since then, the Conservative opposition has toughened its resistance to membership, while the new Labour government has stated that it is in favour of membership in principle, but it has effectively ruled out founder membership.

Even during the most difficult post-Maastricht years, the European Union (EU) (as the 1992 Treaty changed the organisation's name once again) attracted three new members. Austria, Sweden and Finland joined in 1995, although Norway's membership was rejected in a referendum. And just as it looked as if the single currency project had been secured, the European Commission pressed ahead with its proposals for further expansion, this time eastwards. In addition to Cyprus, the European Commission has given the green light for Hungary, Poland, Estonia, the Czech Republic and Slovenia to join the European Union. This expansion of the EU will be particularly challenging because it involves a number of countries that are much poorer than the existing members. At present, Greece is the poorest member of the EU with a per capita GDP of around two-thirds of the EU average. Of the proposed new members from Eastern Europe, the Czech Republic has a GDP per capita at one-half of the EU average, while the others have per capita GDPs in the range 25–40 per cent of the EU average.

Enlargement naturally implies a radical reassessment of the EU's budgetary arrangements. It is clear from the Commission's proposals for budgetary reform that the era of rapid budget growth is over. Although expenditure will continue to grow in real terms (by 17 per cent between 1999 and 2006), it will nevertheless remain constant at around 1.2 per cent of the European Union's GDP up to 2006. The shift in the balance of expenditure from the Common Agricultural Policy to the Structural Actions (i.e. the Structural Funds and the Cohesion Fund) will cease, but the distribution of the funds will change. The prospective member states will receive support from the Structural Actions of up to 4 per cent of their GDPs. Some areas that previously qualified for assistance in the current EU will undoubtedly lose funding, although a transition period is envisaged.

Housing research in Europe

Since the mid-1980s, serious attempts have been undertaken to make housing research more international. Within Europe, two broad traditions have emerged from these ambitions: comparative housing studies on the one hand, and research on the effects of European integration on the other. Both traditions are present in this volume. They can also be found in the literature published to date: variants of comparative housing research, employing different theories and methods, at different levels of analysis; and studies on European integration, mainly commissioned by supranational organisations interested in the housing question.

These traditions have been running parallel for some years now, with a few exceptions mentioned below. We argue for co-operation between comparative and EU-related housing research. European integration has advanced to the point where national housing systems cannot be treated as autonomous units of comparative analysis any longer. This is not because of any prospect of a housing policy being developed at the Union level, but due to the indirect, back-door effects of European integration upon national housing markets and policies.

A new agenda had been set by European policy makers, not by the housing research community itself, referred to by Kemeny (1992: 16) as 'epistemic drift' at the national level. The comparative policy agenda almost disappeared beneath the bulk of European integration-cum-housing studies precipitated by the drive towards European integration. This was helped by the fact that other disciplines and researchers were doing the job: economists, lawyers and European specialists joined the political scientists, sociologists and geographers who had been active in comparative housing research previously. Only some economists have contributed to both sides of the camp (e.g. Ghékiere 1991, 1992; Kleinman 1992, 1996; McCrone and Stephens 1995).

Since Maastricht, comparative housing studies have continued to be published (e.g. Balchin 1996; Barlow and Duncan 1992, 1994; Boelhouwer and van der Heijden 1992; Doling 1997; Forrest and Murie 1995; Hallett 1993; Harloe 1995; Hedman 1994; Karn and Wolman 1992; Kemeny 1995; Kleinman 1996; Oxley and Smith 1996; Padovani 1995; Papa 1992; Power 1993; Rudolph and Cleff 1996; Wiktorin 1993), but in many cases they are being developed in relative isolation from EU-focused housing research, and sometimes from each other too. Hence two types of housing studies may be distinguished: comparative housing studies, both at the international and the inter-regional levels, and studies on the impact of European integration upon housing. A third type of research is only emerging, i.e. a more integrated view of developments at the European, national and regional levels. The following sections discuss the strengths and weaknesses of these three types of international housing research in turn. The next section includes some brief comments as to the contributions of the following chapters towards a more comprehensive, truly European housing research.

Comparative housing research

As is the case with housing research in general, comparative housing research can be described as a *multi*-disciplinary field of studies, not as a field of truly *inter*-disciplinary analysis (Kemeny 1992: 11); and even this is true only for the most theorised attempts at comparison. Many other and most earlier comparative housing studies lack a theoretical and a disciplinary focus at all, at least in an explicit form. By means of implicit and covert statements, many of these studies can still be related to broader traditions in cross-national research, such as convergence theory, and corporatist theory (e.g. Schmidt 1989; Kleinman in this volume). The kind of housing research which is not explicitly related to any discipline, paradigm or theory, was encouraged by the early institutionalisation of housing studies, at least in a number of European countries. For example, building research institutes used to be organised around the lowest common denominator, around bricks and mortar and other purely empirical and policy issues.

In his book on *Housing and Social Theory* (1992), Kemeny proposes a way out of the current isolation of housing studies from contemporary debates in the social sciences. It is a strategy that seems viable also in comparative research on housing: researchers should turn to the disciplinary bases of their kind of (comparative) housing research, and reconceptualise it as part of these separate disciplines. That means, there should be more of a (comparative) housing sociology, more of (comparative) housing economics, more (comparative) housing policy, etc. Only after such a transitional phase of 'firm anchoring in the individual disciplines . . . interactive analysis across disciplines can be developed' (Kemeny 1992: 11–18).

Let us consider one example of what such a re-integration of comparative housing research into a particular discipline would look like, and what theoretical and methodological advances such a reconceptualisation might yield. Political science has been chosen because of its leading role in international comparative research. Any literature search will come out with similar results: apart from *comparative linguistics*, there is an abundance of publications on *comparative government*, *comparative politics* and the like. *Comparative sociology*, on the other hand, seems to be strong in methodological debate (see Ragin 1987; Øyen 1990; Immerfall 1991, 1995; Janoski and Hicks 1994; Pickvance 1995; Inkeles and Sasaki 1996; Crow 1997). There are considerably fewer publications on *comparative economic systems* (e.g. Gregory and Stuart 1995; see also the remark by Rose 1991: 452). Comparative housing studies are even rarer, but what is virtually entirely lacking is the inclusion of housing into a policy comparison of countries and regions. The notable exception to this is Heidenheimer *et al.*'s book on *Comparative Public Policy* (1990).

Comparative housing studies are a Cinderella amongst the comparative social sciences, just as housing itself is disliked by all kinds of social scientists: it is neither exclusively a technical problem, nor is it purely market-supplied, nor is it provided by the welfare state only. Kemeny (1992: 79) quotes Wilensky, the doyen of public policy research, as saying, 'A bewildering array of fiscal, monetary, and other policies that affect housing directly and indirectly – even remotely – have made the task of comparative analysis of public spending in this area nearly impossible.'

Some leading scholars have left the field instead of devoting more energy to an integration of housing into comparative analysis. Within political science, the enormous task of feeding housing issues into the expanding fields of both policy analysis and comparative politics, and their combination into comparative policy analysis, remains to be done (for a review in German see Manfred Schmidt 1991, 1993). In a widely debated paper, Stephan Schmidt (1989), a political scientist, began to relate theories of welfare state activity to housing policies and outcomes. His comparative database would have merited expansion and interpretation by means of policy analysis. Lundqvist (1991), another political scientist, set out a programme for a policy-focused, but not policy-restricted comparative housing research. Housing issues, which had been of prime interest to Dunleavy in his early works (e.g. 1981), receive scant notice in his more recent work (1991: 120ff.). All these scholars have left the field of housing, and they have left substantial gaps in housing and comparative housing research.

There are some more arguments for *the resurrection of policy* as one of the foci of *comparative housing research*, to paraphrase Lundqvist's (1991) title. There is a supranational, European level to be unravelled by policy analysis. So far, political science has concentrated on the legal structures and formal processes of European government, excluding the content and policy making at the European level (Schumann 1993: 394). As will be demonstrated in the next section of this chapter, these policies are almost devoid of any *direct* impact upon housing, but there are a number of other fields of European policy which impinge *indirectly* on housing. Any investigation of how these policies are produced, and of the institutions and policy networks involved, should be of considerable interest to European housing research.

The deepening and widening of the European Union is having another, positive side effect upon comparative housing research that has yet to be fully appreciated. In recent years, the data and information bases on member states have been expanded dramatically and improved by a variety of institutions, EU agencies and interest groupings. For example, statistical series have been published by the European Commission, the Observatory for Social Housing (since 1994) and by the European Mortgage Federation. On a world-wide level, the United Nations' 'Housing Indicators' database is being extended to all countries. These series mark a great improvement on what the UN-ECE's *Annual Bulletin of Housing and Building Statistics* has had to offer since 1957; and even that publication's coverage was enlarged in 1989. However, the most recent statistical series are of variable quality. Many are based on national sources, which gives rise to serious problems of coverage and comparability, severely limits the ability of economists to conduct rigorous statistical analyses (see Ball and Grilli 1997) and has led others to recommend a greater role for the EU in co-ordinating statistical series (Maclennan *et al.* 1997).

European integration and housing research

The European Parliament and various NGOs with an interest in housing have attempted to place housing on the Commission's agenda, so far with little success. The sometimes confusing array of resolutions, communiqués and charters, that has come out of this process is summarised in McCrone and Stephens

(1995). Those who would like to see the European Union take a more active role in housing policy have made the following demands: (1) the right to housing should be included in the Community legislation, particularly in the Social Charter, (2) national housing policies should pay due attention to the applicability of the right to housing, (3) funds should be allocated to EC budget line B3–413, to be used for research and pilot projects, as well as for European housing organisations (UNFOHLM 1992: 7). These demands are found in a document by possibly the strongest lobbyist in the field, the French association of non-profit housing associations, which has initiated several institutions at the European level: Mission Europe and CECODHAS (the European Liaison Committee for Social Housing) in 1988, and the European Social Housing Observation Unit in 1993. Together with six other NGOs in the field (OEIL-JT, FEANTSA, EUROPIL, European Network of Researchers, HIC Europe, AITEC), CECODHAS signed a 'European Charter against Exclusion and for Housing Rights' in 1991.

In the aftermath of the Single European Act of 1987, several resolutions have been adopted by the European Parliament which address the problem of homelessness and relate it to the newly formulated goal of 'economic and social cohesion' (Articles 130A to 130E of the Treaty) that is being endangered by an insufficient supply of affordable housing (see Fitzpatrick in this volume; Daly 1996). As in the European Parliament's resolutions, the member states are exhorted to include a right to housing in the list of fundamental social rights of their citizens. In at least two of its resolutions, the Parliament goes as far as to demand the development of a Community policy in the field of housing. The Commission, on the other hand, is urged to investigate the housing situation across all EU countries, with special regard to the most deprived. The Parliament's most recent resolution (in May 1997) is no exception.

Due to this legacy, and to more recent activities of both the European Parliament and European interest groups, the Commission's view of housing has become strictly geared towards the disadvantaged minority (see Kleinman in this volume). The implicit idea is that if the EU has anything to do with shelter and housing, then it has to relate to economic and social cohesion, and resulting problems be alleviated by means of social and, if possible, regional policy. Other aspects of housing provision, the circumstances of its production, the circumstances of its exchange, are not considered a housing question. The bulk of European legislation is in these spheres, however, and impinges upon the chain of housing provision in many ways (Ambrose 1991; or *filière-logement*, as in Ghékiere 1992: 129). These 'indirect' effects upon housing emanate from the Union's general economic policy. They have become the focus of some European housing research in recent years.

Related to the two important steps towards European integration, the Single European Act and the Maastricht Treaty, two waves of research can be distinguished: the first looks at the completion of the single market, the second evaluates the impacts of monetary union. Both of them are dominated by economists.

Ghékiere (1991, 1992) probably was the first to analyse the bulk of single market directives with regard to their consequences for housing construction, for housing management, for housing transactions, for housing finance, and for the legal status of companies. His findings are summarised in a 'model of convergence' of housing policies. A monetarist ideology of minimising state intervention and a related belief in market efficiency are leading European nation-states to withdraw from housing market intervention, albeit in a variety of ways. Nevertheless, two general processes can be observed: one concerning the instruments of housing policy, the other concerning the actors and institutions involved. The instruments are increasingly targeted towards individual households, whilst housing institutions are becoming less specialised, and increasingly operate at sub-national levels (Ghékiere 1992: 212–219).

For the purposes of British housing associations and local housing authorities, Drake has produced two more practice-oriented reports (1991, 1992). A shorter report was given by Lyons (1992). Reports evaluating the use of European Structural Funds by social landlords have been commissioned by the Housing Corporation and Scottish Homes (Stephens et al. 1996, 1997). Even more targeted on the needs of one particular British housing authority (Cardiff) is the report by Williams and Bridge (1993). The first 'official' summary of EU legislation with regard to housing production was published by DG V (Wyles 1994).

Theoretically, housing finance should be affected first by the completion of the single market, due to capital being more mobile than other factors of production. Hence, a number of research projects have centred on the housing effects of capital market deregulation, both within and beyond the EU (see Bartlett and Bramley 1994). In fact, the relevant directives (Capital Market Directive 88/361/EEC, Second Banking Directive 89/646/EEC) are not only to be seen as instruments of the single market, but also as a reflection of a world-wide trend towards deregulated and competitive capital markets. In Britain, which has been a pioneer in this field, EU policy can be said to be 'pushing against an open door' (Whitehead 1994: 29).

The first systematic evaluation of the Maastricht Treaty with regard to housing has come from members of the European Network for Housing Research. At the request of DG V, Priemus et al. (1993) produced a report on the potential impacts of European monetary and economic union upon national housing policies.

As in the Single European Act before, any reference to housing has been avoided in the Maastricht Treaty (in contrast to town and country planning: see Barlow in this volume). Nevertheless, it is the road towards a single currency which has attracted most interest from housing researchers. Priemus et al. concluded that 'If anything, convergence and union are likely to accelerate some of the trends in housing policies already emerging at the start of the 1990s' (1993: 30).

More recently, the importance of housing systems within a single currency area has been considered. Stephens (1995) focused explicitly on the problems arising from house price volatility in some countries. In a report for the Council of Mortgage Lenders, Maclennan and Stephens (1997) highlighted the importance of the housing system to regional economies adjusting to economic shocks, a theme also developed by Ball and Grilli (1997) in their report to the Royal Institute of Chartered Surveyors.

Both waves of EU-focused housing research have encountered the same difficulties of separating market processes from the effects of supranational policy. This is due to the fact that the Union's primary and secondary legislation are not just instruments by which economic integration is achieved, but that they are also a reflection of market-led developments. It is quite difficult, therefore, to disentangle global-market-induced and EU-policy-induced changes of European integration. For practical reasons, such a distinction was made by Priemus *et al.* (1993: 34).

The structure of this book

Part I of this book is devoted to markets, with contributions relating both to changes in the global economy to which European national economies have reacted, and the impact of European integration itself. Christine Whitehead sets the stage with a review of integration in capital markets. She finds that despite capital market liberalisation, the distinctive characteristics of mortgage systems have persisted at the national level. Michael Oxley and Jacqueline Smith search for the determinants of housing investment levels amongst EU member states. Michael Ball and Michael Harloe follow with a review of recent changes in housing provision, deliberately including the USA, and downplaying any specific impact of European integration upon housing. Hugo Priemus concludes with an evaluation of the direct and indirect consequences of EU legislation for the construction industry.

Even in Priemus's contribution, where the supranational level of policy is most explicitly researched, its impact upon national and regional markets is found to be limited, and divergent organisational patterns of the building process are predicted to persist. Convergence is seen to be driven by more general, global processes, which are not always admitted by national and/or European policies. On such occasions the territorial or scale dimension of the convergence/divergence debate comes to the fore: there may be convergence amongst member states within the EU, but divergence may continue or aggravate between regions, sectors and classes of their economies and societies.

In Part II, the focus shifts towards policies at the national and regional levels that are being affected by policy changes at the European level. Chapters in this section tend to concede a somewhat greater role to the Union in shaping national outcomes. The interplay between different levels and sectors of European policy receive more attention than used to be the case. In places the task can only be achieved by means of reducing the comparative database to a number of case studies.

Mark Stephens analyses the impact of fiscal restraint upon public housing expenditure, currently the strongest effect that European integration exerts upon national housing policies. James Barlow continues with a review of recent EU involvement in regional planning and environmental issues. This is an area which used to be dominated by the legal and institutional legacies of nation states. Recently, a number of comparative studies have extended our knowledge of different types of urban planning (see Newman and Thornley 1996) and different trends in regional development within Europe (OOPEC 1996). These studies will not have any immediate repercussions in EU legislation, but they will continue to inform the policy agenda at the national and European levels (Kunzmann 1996). Valerie Karn and Louise Nyström discuss another neglected aspect of housing policy, the setting of building standards and recent trends towards their deregulation within the EU and beyond.

Part III is devoted to the social outcomes of housing markets and policies across Europe. Chapters range from the 'true' housing costs of homeowners to the problems of the socially excluded. Again these contributions are more on the comparative side of European housing research, and references to the role of the Union are limited, so emphasising its currently limited direct role.

To date, the European Union does not interfere in direct taxation (with a few exceptions, mainly deriving from European Court decisions). With regard to homeowners, the existing variety of national systems of taxation is amply demonstrated in Marietta Haffner's chapter. It is her stated goal to make these different ways of taxation and subsidisation comparable, by means of applying the concept of user costs (instead of using the standard method of housing expenses). John Hills takes an even wider view on the impact of both owning and renting upon the distribution of incomes. Any further policy at the Union level should be informed by the results of such advances in comparative methodology. It remains to be seen whether and when future amendments to the Treaty will include guidelines as to the definition of taxable income. With regard to corporate income taxation, such a harmonisation of the assessment basis has already been proposed by the Ruding Committee Report (CEC 1992).

Suzanne Fitzpatrick analyses the causes, characteristics and extent of homelessness in the European Union, and considers whether the very limited actions of the EU in this area should be extended further. Anne Power makes accessible yet another persistent problem of European housing, that of run-down housing estates. Well before recent moves towards European integration, in the 1960s and 1970s, each country had developed its variant of mass housing. Comparative analysis of selected case studies sheds new light on similarities and differences in housing management and rehabilitation.

Trying to avoid the traditional schism between convergence and divergence approaches, Mark Kleinman proposes the notion of 'policy collapse' as being the main characteristic of Western European housing policies. Over the last twenty years, responsibilities have been devolved down to sub-national levels, and housing policy has become fragmented into issues of poverty and social exclusion on

11

the one hand, and into issues related to safeguarding the housing situation of the better off on the other hand. There is considerable agreement here with other comparative findings discussed above, and the related observation that, if any, only the poverty-related issues of housing are being considered by the Union and its institutions.

Part IV looks at the current patterns of diversity amongst national housing policies and their further development. Sirpa Tulla adds information on the housing situation within the Nordic countries, inside and outside the EU. Despite decades of social policy harmonisation, housing markets and policies have remained surprisingly different in these countries, and EU membership *per se* has hardly made any difference so far. Since the collapse of communism, housing subsidies and new construction have been dramatically reduced, and a substantial part of the stock has now been privatised. Iván Tosics shows the diversity that exists in housing amongst the transitional economies in Central Eastern, South Eastern and North Eastern Europe. Compared to other aspects of an impending membership of the EU (at least for the first group of countries) housing will not be such a problem. As within the existing Union, the consequences of integration will be indirect, consisting of mainly side effects arising from economic legislation, plus a few direct social policy programmes for the deprived. Perhaps of greater importance will be the effects of integrating new candidates into the circuits of information exchange that span the Union: pilot, demonstration and 'best practice' projects are deliberately aimed at imitation, far beyond the places and institutions involved. By means of comparative statistics and reports, investment opportunities will be recognised by builders, developers and lenders, and the virtues of maintaining some efficient form of social housing will become visible to East European politicians and administrators.

Across all chapters of this volume, a multiplicity of relations between comparative housing research and EU-focused housing research is being addressed. Nevertheless, all chapters can still be classified as either being mainly comparative or mainly centred on the EU. The two strands of housing research coexist, and this is to be welcomed, 'the relation between European integration and national housing policies [being] bilateral' (Priemus *et al.* 1993: 53). The two traditions should grow together in the years to come.

References

Ambrose, P. (1991) 'The Housing Provision Chain as a Comparative Analytical Framework', *Scandinavian Housing & Planning Research* 8, 2: 91–104.

Balchin, P. (ed.)(1996) *Housing Policy in Europe.* London: Routledge.

Ball, M. and Grilli, M. (1997) *Housing Markets and Economic Convergence in the European Union.* London: Royal Institution of Chartered Surveyors.

Barlow, J. (1998) 'Planning, Housing and the European Union', in this volume.

Barlow, J. and Duncan, S. (1992) *Markets, States and Housing Provision: Four European Growth Regions Compared.* Oxford: Pergamon Press.

Barlow, J. and Duncan, S. (1994) *Success and Failure in Housing Provision. European Systems Compared*. Oxford: Pergamon (Elsevier Science).

Bartlett, W. and Bramley, G. (eds) (1994) *European Housing Finance. Single Market or Mosaic?* Bristol: SAUS Publications.

Boelhouwer, P. and van der Heijden, H. (1992) *Housing Systems in Europe: Part I. A Comparative Study of Housing Policy* (Housing and Urban Policy Studies 1). Delft: Delft University Press.

CEC (Commission of the European Communities) (1992) *Conclusions and Recommendations of the Committee of Independent Experts on Company Taxation* (Ruding Committee). Luxembourg: Office for Official Publications of the European Communities.

Crow, G. (1997) *Comparative Sociology and Social Theory. Beyond the Three Worlds*. Houndmills and London: Macmillan.

Daly, G. (1996) *Homeless. Policies, Strategies, and Lives on the Street*. London: Routledge.

Doling, J. (1997) *Comparative Housing Policy: Government and Housing in Advanced Industrialized Countries*. Basingstoke: Macmillan.

Drake, M. (1991) *Housing Associations and 1992: The Impact of the Single European Market*. London: National Federation of Housing Associations and the Housing Corporation.

Drake, M. (1992) *Europe and 1992. A Handbook for Local Housing Authorities*. Coventry: The Institute of Housing.

Dunleavy, P. (1981) *The Politics of Mass Housing in Britain, 1945–1975. A Study of Corporate Power and Professional Influence in the Welfare State*. Oxford: Clarendon Press.

Dunleavy, P. (1991) *Democracy, Bureaucracy and Public Choice*. New York: Harvester-Wheatsheaf.

Fitzpatrick, S. (1998) 'Homelessness in the European Union', in this volume.

Forrest, R. and Murie, A. (eds) (1995) *Housing and Family Wealth. Comparative International Perspectives*. London: Routledge.

Ghékiere, L. (1991) *Marchés et politiques du logement dans la CEE*. Paris: Documentation Française.

Ghékiere, L. (1992) *Les Politiques du logement dans l'Europe de demain*. Paris: Documentation Française.

Gregory, P.R. and Stuart, R.C. (1995) *Comparative Economic Systems*. Boston, Toronto: Houghton Mifflin.

Hallett, G. (ed.) (1993) *The New Housing Shortage. Housing Affordability in Europe and the USA*. London: Routledge.

Harloe, M. (1995) *The People's Home? Social Rented Housing in Europe and America*. Oxford: Blackwell.

Hedman, E. (ed.) (1994) *Housing in Sweden in an International Perspective* (Boende Rapport 1993: 2 e). Karlskrona: Boverket.

Heidenheimer, A., Heclo, H. and Adams, C. (1990) *Comparative Public Policy. The Politics of Social Choice in Europe and America*, 3rd edn. New York: St Martin's Press. (1st edn 1975, 2nd edn 1983.)

Immerfall, S. (1991) 'Der Vergleich als Methode der empirischen Sozialforschung. Anmerkungen zu Status und Strategien vergleichender Vorgehensweisen', *SWS-Rundschau*, 31, 4: 551–568.

Immerfall, S. (1995) *Einführung in den europäischen Gesellschaftsvergleich. Ansätze – Problemstellungen – Befunde*. Passau: Wissenschaftsverlag Richard Rothe.

Inkeles, A. and Sasaki, M. (eds) (1996) *Comparing Nations and Cultures. Readings in a Cross-Disciplinary Perspective*. Englewood Cliffs, NJ: Prentice-Hall.

Janoski, T. and Hicks, A.M. (1994) *The Comparative Political Economy of the Welfare State* (Cambridge Studies in Comparative Politics). Cambridge: Cambridge University Press.

Karn, V. and Wolman, H. (1992) *Comparing Housing Systems: Housing Performance and Housing Policy in the United States and Britain*. Oxford: Oxford University Press.

Kemeny, J. (1992) *Housing and Social Theory*. London: Routledge.

Kemeny, J. (1995) *From Public Housing to the Social Market. Rental Policy Strategies in Comparative Perspective*. London: Routledge.

Kleinman, M. (1992) *Policy Responses to Changing Housing Markets: Towards a European Housing Policy?* (Welfare State Programme, Discussion Paper Number WSP/73). London: STICERD, LSE.

Kleinman, M. (1996) *Housing, Welfare and the State in Europe. A Comparative Analysis of Britain, France and Germany*. Cheltenham: Edward Elgar.

Kleinman, M. (1998) 'Western European Housing Policies: Convergence or Collapse?', in this volume.

Kunzmann, K. (1996) 'Euro-megalopolis or Themepark Europe? Scenarios for European Spatial Development', *International Planning Studies*, 1, 2: 143–163.

Lundqvist, L. (1991) 'Rolling Stones for the Resurrection of Policy as the Focus of Comparative Housing Research', Special Issue on Comparative Housing Research, *Scandinavian Housing & Planning Research*, 8, 2: 79–90.

Lyons, R. (1992) *The Implications of European Social and Economic Integration for Housing in the United Kingdom* (School of Land Management and Urban Policy, Occasional Paper 3/1992). London: Faculty of the Built Environment, South Bank University.

Maclennan, D. and Stephens, M. (1997) *EMU and the UK Housing and Mortgage Markets*, London: Council of Mortgage Lenders.

Maclennan, D., Stephens, M. and Kemp, P. (1997) *Housing Policy in the EU Member States*. Luxembourg: European Parliament Directorate General for Research.

McCrone, G. and Stephens, M. (1995) *Housing Policy in Britain and Europe*. London: UCL Press.

Newman, P. and Thornley, A. (1996) *Urban Planning in Europe. International Competition, National Systems and Planning Projects*. London: Routledge.

OOPEC (1996) *First Report on Economic and Social Cohesion 1996* (CM-97-96-928-EN-C). Luxembourg: Office for Official Publications of the European Communities.

Oxley, M. and Smith, J. (1996) *Housing Policy and Rented Housing in Europe*. London: E. and F.N. Spon.

Øyen, E. (ed.) (1990) *Comparative Methodology. Theory and Practice in International Social Research* (SAGE Studies in International Sociology, vol. 40). London: Sage.

Padovani, L. (ed.) (1995) *Urban Change and Housing Policies. Evidence from Four European Countries* (DAEST, Collana Ricerca n. 19). Venice: Istituto Universitario di Architettura di Venezia.

Papa, O. (1992) *Housing Systems in Europe: Part II. A Comparative Study of Housing Finance* (Housing and Urban Policy Studies 2). Delft: Delft University Press.

Pickvance, C. (1995) 'Comparative Analysis, Causality and Case Studies in Urban Studies', pp. 35–54 in A. Rogers and S. Vertovec (eds) *The Urban Context. Ethnicity, Social Networks and Situational Analysis.* Oxford: Berg Publishers.

Power, A. (1993) *Hovels to Highrise: State Housing in Europe since 1850.* Routledge: London.

Priemus, H., Kleinman, M., Maclennan, D. and Turner, B. (1993) *European Monetary, Economic and Political Union: Consequences for National Housing Policies* (Housing and Urban Policy Studies 6). Delft: Delft University Press.

Ragin, C. (1987) *The Comparative Method. Moving Beyond Qualitative and Quantitative Strategies.* Berkeley, Los Angeles and London: University of California Press.

Rose, R. (1991) 'Comparing Forms of Comparative Analysis', *Political Studies,* 39, 3: 446–462.

Rudolph and Cleff (1996) *Wohnungspolitik und Stadtentwicklung. Ein deutsch–französischer Vergleich* (Stadtforschung aktuell, vol. 55). Basel, Boston, Berlin: Birkhäuser.

Schmidt, M. (1991) 'Vergleichende Policy-Forschung', pp. 197–212 in D. Berg-Schlosser, and F. Müller-Rommel (eds) *Vergleichende Politikwissenschaft* (UTB 1391). Opladen: Leske+Budrich.

Schmidt, M. (1993) 'Theorien in der international vergleichenden Staatstätigkeitsforschung', pp. 371–393 in A. Héritier (ed.) *Policy-Analyse. Kritik und Neuorientierung* (Politische Vierteljahresschrift, Sonderheft 24/1993). Opladen: Westdeutscher Verlag.

Schmidt, S. (1989) 'Convergence Theory, Labour Movements, and Corporatism: The Case of Housing', *Scandinavian Housing & Planning Research,* 6, 2: 83–101.

Schumann, W. (1993) 'Die EG als neuer Anwendungsbereich für die Policy-Analyse: Möglichkeiten und Perspektiven der konzeptionellen Weiterentwicklung', pp. 394–431 in A. Héritier (ed.) *Policy-Analyse. Kritik und Neuorientierung* (Politische Vierteljahresschrift, Sonderheft 24/1993). Opladen: Westdeutscher Verlag.

Stephens, M. (1995) 'Monetary Policy and House Price Volatility in Western Europe', *Housing Studies* 10: 551–564.

Stephens, M., Bennett, A. and Smith, F. (1996) *European Funding: The Independent Social Housing Sector and the European Structural Funds in England.* London: National Housing Federation.

Stephens, M., Bennett, A., and Smith, F. (1997) *Housing Associations and the European Structural Funds in Scotland,* Edinburgh: Scottish Homes.

UN-ECE (annually, since 1958) *Annual Bulletin of Housing and Building Statistics for Europe.* Geneva: United Nations.

UNFOHLM (ed.) (1992) *Rapport du groupe interfédéral Europe. 53e Congrès National HLM Strasbourg 11/15 juin 1992.* Paris: Union Nationale des Fédérations d'Organismes HLM.

Whitehead, C. (1994) 'The Opening Up of UK Housing Finance', pp. 19–45 in W. Bartlett and G. Bramley (eds) *European Housing Finance. Single Market or Mosaic?* Bristol: SAUS Publications.

Wiktorin, M. (1993) *An International Comparison of Rent Setting and Conflict Resolution.* Gävle: Swedish Institute for Building Research.

Williams, P. and Bridge, G. (1993) *Cardiff City into Europe. The Impact of the Single European Market on Cardiff's Housing Service* (Papers in Housing Research 6). Cardiff: University of Wales, Cardiff, Department of City and Regional Planning.

Wyles, R. (1994) *Rechtsanalyse zu den Auswirkungen der EG-Gesetzgebung im Wohnungsbau* (Series: Housing in Europe). Brussels: CEC, DG V.

Part I

MARKETS

2

ARE HOUSING FINANCE SYSTEMS CONVERGING WITHIN THE EUROPEAN UNION?

Christine Whitehead

Traditionally, housing finance has been particularly dependent on the individual attributes of each country's financial, legal and policy systems. The outcome has been that different countries tend to have quite different patterns of ownership, property rights and funding – in terms of the institutions involved, the instruments employed, the legal positions of owner, occupier and financier and the importance of government with respect to taxation, subsidy and regulation.

Housing provision and policy is in the main outside the remit of the European Union, because it is a local asset traded in nationally based markets. However, the provision of housing and its finance does have direct impacts on a country's relative trading position. Moreover, housing finance is more and more an intrinsic part of the overall finance system and thus of the single market.

The question of potential harmonisation of housing finance was examined by the European Community in relation to a draft directive on mortgage credit in 1984 (Commission of the European Communities 1984). A report by the House of Lords Select Committee at that time concluded that harmonisation and mutual recognition of finance institutions and indeed savings and mortgage instruments were only a small element in the overall question of effective competition. Housing and its finance were seen to be fundamentally grounded in the different nature of legal systems and institutional structures as much as in taxation and subsidy policies or the right to trade and operate across borders. Without greater convergence across the whole range of factors, they argued, housing finance systems would be likely to remain very separate and to continue to develop in different ways (House of Lords, Select Committee on the European Communities 1985). In the end the draft directive came to nothing and the whole question of housing finance became subsumed into the wider question of banking regulation, which was itself addressed through the Second Banking Directive in 1989 (Commission of the European Communities 1989; British Bankers' Association 1990).

In the context of banking regulation commentators have argued that what liberalisation and adjustment there has been with respect to housing finance has been more the outcome of pressures arising from the globalisation of finance markets than specifically the result of European initiatives. European directives, while stressing cross-border harmonisation, have followed rather than led wider international agreements, which have put far greater stress on free competition (Bartlett and Bramley 1994; Basle Committee of Banking Supervisors 1988).

Now that the Union has technically completed the single market and is moving towards a single currency, it is relevant to revisit the question of convergence and to ask whether there are signs that housing finance systems in individual countries are coming to operate in similar ways – whether as a result of the movement towards European union or wider pressures towards liberalisation and, indeed, reducing public expenditure and government involvement in housing. Related questions include whether or not there is evidence of convergence in institutions and instruments or indeed in outcomes, in particular whether costs and availability are becoming more similar across countries. In this chapter we bring together some evidence on comparative trends, looking at some eleven European Union countries together with Norway, which has been collected for a range of projects, to examine certain aspects of these questions (Whitehead 1996; Freeman *et al.* 1996a; Turner *et al.* 1996).[1] In particular, we look at trends in tenure structure, the institutions and instruments used to provide funding in the private sector, the legal framework for enforcing financing contracts, and the role of government in financing housing.

Tenure patterns

Table 2.1 shows the proportion of households (or in some cases dwellings) in owner-occupation. Much of this information is well known. Two main groups of countries emerge – Germany, Austria, France and the Netherlands with relatively low proportions and the Anglo-Saxon based and southern European countries with a relatively high proportion. The proportions in the Scandinavian countries, which are often thought to be homogeneous, are surprisingly variable.

What is more revealing is the extent of consistency in expectations about changes in that proportion. Only in Greece is owner-occupation predicted to continue to rise rapidly, in the face of very limited alternatives. In Germany and the Netherlands the trend is also upwards but slowly and from a low base. Elsewhere, whatever the existing levels, the proportion is expected, at the most, to remain stable. Thus, in the majority of countries the level of owner-occupation is thought to have reached a plateau, or even to be tending to decline. This does not mean that owner-occupation is becoming less desirable. Indeed, if we look at rather different evidence about the behaviour of average households we find that much higher proportions, ranging from 68 per cent in the Netherlands, and 77 per cent in Germany to 92 per cent in Ireland choose to be owner-occupiers, suggesting that it is still the preferred tenure for those with adequate income to

Table 2.1 Tenure structure (% of households)

Country	Owner-occupation (%)	Expected changes in owner-occupation	Social renting (%)	Private renting (and other) (%)
Finland	78 (1992)	Falling	11	11
Sweden	55 (1991)	Falling slightly	22	23
Norway	78 (1990)	Stable	5	17
Denmark	66 (1994)	Falling	19	15
Netherlands	48 (1993/4)	Rising slowly	40	12
Germany	38 (1987)	Rising slowly	15	47
Austria	55 (1991)	Stable	22	23
Ireland	77 (1987)	Stable	14	9
UK	68 (1994/5)	Stable	23	10
France	54 (1990)	Stable	15	31
Spain	78 (1989)	Stable	1	21
Greece	77 (1994)	Rising		

Sources: Questionnaires to country experts (see note 1).

achieve it comfortably (Freeman *et al.* 1996b). What it may suggest, however, is that deregulation in both housing and finance markets has helped to generate a wider range of housing choice.

When we look at the rented sectors we again find large variations with respect to the relative importance of social and private provision. The Netherlands, the UK and Sweden stand out, with large social sectors, while Germany, France and much of Scandanavia have large private sectors. These differences can be traced in part to the different forms of intervention and subsidy that have been available, with Scandinavia and Germany providing more investment-based but tenure-neutral assistance and the UK and the Netherlands historically favouring social provision. Again, however, there is some evidence of consistency with respect to the direction of change in the make-up of the rented sector, with generally greater emphasis on private provision than in the past.

Sources of housing finance for owner-occupation

It has often been predicted that globalisation in finance markets in general, liber-alisation in housing finance in particular, and the completion of the European internal market would all tend to result in standardisation of financial institutions and mortgage instruments (Miles 1994). Tables 2.2 and 2.3 show that, even if this is going to occur, it certainly has not done so yet.

In part the differences are ones of nomenclature: for instance is a mortgage bank distinct from a housing bank in any other sense than the first covers a wider range of land-based assets than the second?

Historically there have been two big substantive differences and one great similarity. The differences concern whether funding for private housing has, on the one hand, been through direct retail institutions or, on the other, through, often state-owned, intermediaries providing matching bonds. Second, but strongly linked to the first, is the question whether government provides direct assistance, usually in the form of interest subsidies, often tied to new provision or investment in housing. The similarity has been in the extent to which there has

Table 2.2 Sources of housing finance

	Main source of funds by tenure
Finland	Private: retail banks
	Social: government
Sweden	All sectors: mortgage banks
Norway	All sectors: housing banks for construction
	Private: retail banks
Denmark	All sectors: mortgage banks
Netherlands	Private: retail banks, insurance companies, mortgage banks
	Social: same, with government guarantees
Germany	Private: *Bausparkassen*, savings banks, mortgage banks
	Social: banks, state
Austria	Private: *Bausparkassen*, retail banks and state
	Social: banks, state
Ireland	Private: retail banks, building societies
	Social: state
UK	Private: building societies, retail banks, some secondary market
	Social: state, private
France	Private: *Crédit Foncier*; PAPs; Caisse d'Epargne; savings plans
	Social: Caisse des Dépôts et des Consignations (CDCs)
Spain	Private: retail banks, savings banks, mortgage banks
	Social: state
Greece	Private: own funds; specialist lenders, retail banks
	Social: state

Sources: Questionnaires to country experts (see note 1).

Table 2.3 Form of mortgage

	Down payment (% of total price)	Length of mortgage	Type of mortgage
Finland	30	7–15 years	Variable
Sweden	5+	20–30 years	Fixed; variable
Norway	30 (but minus 10–25 in boom)	20–30 years	Variable; short-term fixed increasing
Denmark	5+	20–30 years	Fixed only
Netherland	25–minus 25	30 years	Fixed 5–20 years
Germany	20–30	20–30 years	Fixed
Austria	20–30	25 years	Banks: variable; *Bausparkassen*: fixed
Ireland	10	20–30 years	Variable; short-term fixed growing
UK	5–10 (but minus 10–25 in boom)	20–25 years	Variable; short-term fixed
France	10 if contract savings; more if no savings scheme	20 years	Fixed; variable increasing
Spain	20	15 years	Fixed; variable increasing
Greece	25–50	6–15 years	Fixed

Sources: Questionnaires to country experts (see note 1).

been a special circuit of housing finance which has directed funds towards housing investment or transactions and provided them at below market rates through regulation, tax benefits and sometimes subsidy. In many ways it is this similarity that has been disappearing over the last decade, as much as the differences.

In all European countries there has been emphasis on deregulation of the financial system overall which has in turn entailed some integration of housing finance into the general market. There has also in most countries been greater emphasis on privatisation including, where relevant, changing the status of state-owned banks. So far, however, the outcome has been quite different between countries with, at one extreme, notably in France, Germany and particularly Austria, the continuation of a special circuit of housing finance, still usually linked to government assistance and/or guarantee, and at the other, notably in the UK, Spain and other countries that tend more to utilise retail banks, an open finance system almost entirely non-specific to housing. So instead of all countries having a special circuit there are now very real differences in the extent to which housing finance has been integrated into the overall finance market.

Looking next at the differences: in all the countries that we examined that had used interest rate subsidies to encourage investment there have been moves to reduce the extent of interest rate subsidies, either to phase them out altogether or to limit them to particular forms of investment (such as rehabilitation) or particular locations (such as urban regeneration areas). In some cases these benefits have been restricted to rented and particularly social rented housing. At the same time in almost all the countries that had provided mortgage tax relief for borrowing for owner-occupation that relief has either been abolished or restricted. The most important exception here has been the Netherlands which continues to provide full deductibility at the marginal tax rate in all sectors, while at the same time taxing imputed rents (Table 2.4). The tendency towards convergence that has emerged here is that, for tax purposes, the majority of countries now treat owner-occupied housing as a consumption good, while treating rented housing as an investment good.

An important aspect of mortgage funding in many countries has been the role of contract saving. Traditionally, specialist institutions have provided lower rates of interest to small, usually younger, savers in return for access to below market

Table 2.4 Government support to owner-occupation

	General	*Targeted*	*Change in level of support*
Finland	Tax relief at 28% (30% first-time buyers)	Young first-time buyers; contractual savings	Declining
Sweden	Tax relief at 30%; interest subsidies being phased out	New; improvement investment	Declining
Norway	Tax relief at 28%; low imputed income tax	Housing allowances; contractual savings	Declining
Denmark	Tax relief (reducing); low imputed income tax	None	Declining
Netherlands	Tax relief at marginal tax rate	Low income earners before 1993	Declining
Germany	Interest rate subsidies; depreciation allowances	Tax credit for families; improvement	Stable
Austria	Interest rate subsidies; no tax relief	Low income; young; investment	Stable
Ireland	Tax reliefs (declining rapidly from 48%)	New investment	Declining higher incomes; stable lower incomes
UK	Tax reliefs (declining rapidly, now 15%)	Low income (very limited improvement investment)	Declining
France	Tax reliefs (restricted); interest rate subsidies	Contractual savings	Declining
Spain	Tax reliefs	Low income, young, first-time buyers; new/improved – 80% of total, so not really targeted?	Declining

Sources: Questionnaires to country experts (see note 1).

priced mortgage funds – with or without an additional subsidy from government. In Britain and other countries where there has been a strong emphasis on deregulation this special circuit has disappeared, generating a much wider range of options for the saver and market-priced mortgages for the borrower. This tends to result in earlier owner-occupation and higher levels of borrowing as compared to the previous constrained systems. In Germany and Austria, on the other hand, the *Bausparkassen* remain central to the mortgage system, although the proportion of finance provided through contract savings has fallen considerably in the last few years, in the face of financial liberalisation. The same is true of France, where the form of finance is tightly linked to the extent of assistance provided by government which, in turn, tends to be more strongly income related than in the past.

Both Norway and Finland have lately introduced contract savings schemes within a liberalised finance market, to assist entry into owner-occupation for marginal purchasers. In this context contract savings have three distinct roles: to signal the household's preparedness to meet its contractual obligations, to reduce the financial risks involved in providing funds to marginal buyers and to target assistance through a form of matching grant. It remains to be seen whether this revival in contract savings is sustainable in the face of competition from institutions prepared to provide higher loan-to-value ratios, unless large-scale government assistance is also on offer. Once again, it reflects in some ways a growing divergence in approach – but linked to different means of targeting government assistance, rather than to market factors.

The evidence on the forms of mortgage instrument available (Table 2.3) also suggests that convergence is not yet the norm. This was an area specifically discussed in the draft directive (Commission of the European Communities 1984) because some countries had legislation making certain types of mortgage which formed the basis of other countries' systems illegal. For instance those countries that favour fixed rates often limited the use of variable rate mortgages while linkages to insurance, for instance in the form of endowment mortgages, were differently regulated or even outlawed. Harmonisation and mutual recognition have significantly widened the potential range of instruments, while cross-border establishment and trade have allowed mortgage institutions to develop business in other European countries (Bowen 1994). Even so, many differences remain in what is provided, particularly with respect to the type of interest rate. In some countries where fixed rate mortgages have been the norm variable rate mortgages are becoming available, and vice versa – although in traditionally variable rate systems the duration of the fixing being introduced is still relatively short. Again this is in part the outcome of history – countries with matching funding arrangements clearly find it easier to provide longer-term fixed interest finance. In principle, cross-border competition could address this issue. At the present time, exchange rate risks probably continue to outweigh interest rate risks, so the potential for integration has yet to be fully realised.

Low down payments similarly appear to be associated with the existence of

25

mortgage banks, as do long maturities. Retail bank dominated systems have tended in the past to provide shorter maturities. In the more fully deregulated systems such as the UK, lower down payments, linked with insurance have become an important aspect of the market. In this way different methods of financing together with increased competition have produced more similar outcomes. However, only in the Netherlands is there consistent evidence of borrowing above 100 per cent for certain low-risk households, although institutions in a number of other countries lent over 100 per cent during the boom years of the late 1980s. More generally, the evidence is that in most countries the range of instruments available has increased – but that change is slow.

A relevant question is whether increases in the range of providers and in the availability of funds will tend to modify levels of owner-occupation. Greece is an example where owner-occupation has thrived without significant formal debt finance (as indeed is Italy). Otherwise high proportions of owner-occupation do tend to be related to a wider range of sources, including, in particular, retail banks. Specialist financial institutions tend to go with lower proportions of individual ownership. Again we can query whether this will change as competition grows or whether the differences are mainly the outcome of other factors.

What we currently observe, therefore, is in some ways greater differences between countries in the extent to which funding is part of a special circuit, although within a more liberal overall finance system. This may anyway turn out to be only a transitional phase, in that, as savings opportunities and competition for lending both increase and especially as government assistance declines, survival will come to depend more and more on efficiency. This does not mean that institutions or even instruments will be standardised across countries – but it should mean that savers, mortgagors and institutions in these countries face similar opportunities.

The legal framework for arrears and foreclosures

One of the most important factors relevant to convergence is the operation of the legal system when it comes to enforcement of contracts – for this helps to determine what types of contract can be effectively issued, and may mean that while a particular type of instrument can operate effectively in one country it will be unsuitable for conditions in another. An extreme example was where a particular form of contract, such as a variable rate mortgage, was illegal in a particular country so the terms and conditions could not be enforced. A more general problem relates to the capacity of the mortgagee to recover mortgage arrears or to gain possession of the property where payment is not forthcoming. These procedures for dealing with arrears and foreclosures help to define the relative bargaining position of lenders and borrowers in the event of unexpected difficulties. They are thus fundamental to decisions about the types of household that can obtain a mortgage, on what terms and for what types of dwelling.

Table 2.5 provides some limited detail on the position in different countries.

Table 2.5 Arrears and foreclosures

	When is mortgage defined as in arrears?	When is action taken?	What form?
Finland	90 days	No rules	Court order/ restructuring
Sweden	6 months	3–6 months	Talk/court order/ restructuring
Norway	3 months	Immediate	Restructuring/ court order
Denmark	1 month	6–12 months	Foreclosure
Netherlands	4 months	Immediate	Foreclosure
Germany	Negligible problem	Mechanisms hardly used	
Austria	Negligible problem	Mechanisms hardly used	
Ireland	Immediately	Immediately	10–15 steps to foreclosure – almost never occurs
UK	Immediately	No rules	Restructuring/ foreclosure
France	Not known		
Spain	90 days	2–4 months	Entity/notary/ court
Greece	1 year	1 year	Court but possession rarely given

Sources: Questionnaires to country experts (see note 1).

In more traditional regimes, with special circuits of housing finance, the problem of arrears and foreclosures has been dealt with mainly by simply not lending anywhere near the margin. This is still the case in Germany and Austria in particular, where arrears and foreclosures are not seen as a relevant question. Where the problem is seen as of particular relevance is in countries like Finland, and to a lesser extent Sweden, with higher proportions of owner-occupiers, but where, like Germany, there is no general legislation allowing individual bankruptcy. In the context of housing market recession and sudden changes in interest rates and individual economic circumstance, this has led to the development of special government assisted schemes for restructuring and writing-off debt.

In the more market-oriented systems of the United States, Australia and Britain, on the other hand, there are general bankruptcy and foreclosure laws which are applied to housing in the same way as to other assets. Even so, housing is still regarded as special, in part because foreclosure is seen as a social problem and in part because the process is very expensive and therefore, if possible, should be avoided from the point of view of borrowers and lenders alike.

Perhaps for this reason, whatever the formal rules, restructuring debt is by far the most usual practice in the face of growing arrears. In some countries this is forced upon institutions because the formal enforcement powers are not applied by the courts and it is therefore almost impossible to obtain possession. In others it is more a business decision based on the relative costs of different approaches. In either case these conditions affect the ways in which contracts operate and therefore the costs of borrowing.

Higher proportions of owner-occupiers and more liberal finance markets have undoubtedly brought the question of arrears and foreclosure to the forefront of discussion – although the upswing in most economies observed in the mid-1990s has reduced the immediacy of concerns. What is also clear is that so far these arrangements are seen as entirely a national concern and there is no sign of any formal convergence in approach.

Financing social housing

The provision and financing of social housing is also of relevance to the question of convergence in that it provides different opportunities. So far, however, the methods by which social housing is funded are seen as aspects of national policies of little direct relevance to the question of convergence. Even so it is worth noting some consistent trends.

The majority of social housing finance is still provided by the state, or involves important government guarantees (Table 2.2). There have, however, been very significant changes in the ways that governments assist provision. In particular, there has been a growth in the use of arm's-length organisations, to the point where some countries have no municipally owned housing at all. Most such organisations remain locally based and continue to have strong links with the local authority; sometimes, as in the Netherlands and France, involving guarantees by that authority – but the finance, excluding subsidy, comes from the private sector. In the UK new social housing provision has shifted from local authorities to housing associations, involving increasing proportions of private funding. In addition there is a growing emphasis on transferring existing stock to associations or local housing companies through privately funded buy-outs. In France there have been fewer structural changes, in part because social housing has always been more flexibly structured with organisations sometimes being controlled by municipalities and sometimes by independent bodies. Germany is once again atypical in defining social housing not by ownership but by the nature of the subsidy provided together with the contract between the relevant landlord and the government. Private landlords may provide social housing for the period of subsidy, usually thirty years, and are then able to let the accommodation on the open market. Austria, while operating in a way similar to Germany, maintains stronger government involvement.

The tendencies across Europe away from both general and open-ended subsidies towards both targeting and certainty of commitment are clear (Turner *et al.* 1996). The Netherlands has gone furthest by rolling up all existing open-ended subsidies into a single grant – together with insurance and guarantees. A number of countries are beginning to limit interest rate subsidies to improvement investment instead of new building: Denmark and France are good examples here. Up-front grants are also becoming more usual, especially for rehabilitation. Britain has been in the forefront of both cash-limiting and locational targeting, with housing associations bidding for grants on the basis of cost as well as need. Thus the emphasis has moved towards greater integration of social housing into the private finance system as well as increasing restrictions on the extent of subsidy. In this context there have certainly been elements of convergence in policy terms.

The other major tendency has been towards person-related, rather than dwelling-related, subsidies. In some countries these are available in all tenures; in others, such as the Netherlands, the UK and Denmark, only in the rented sector. The forms of this assistance vary greatly between countries, but the move towards assisting lower income households rather than directly encouraging the supply side is general. In principle this can allow rents better to reflect the resource costs involved, but there is a long way to go before this becomes the norm.

Overall there is some evidence of convergence with respect to social housing policies and in terms of the mechanisms used for assisting lower income households to obtain adequate accommodation. These tendencies do not arise specifically from the development of the Union, but more from wider pressures on governments to control their expenditures and to target what assistance remains towards particular groups and locations. This second strand of policy has more to do with regional than financial policy, but is no less important in the longer term.

Conclusions

The extent of convergence in terms of financial institutions and instruments remains relatively limited. What matters, however, is not whether each country operates in the same way, but whether households and providers in each country have similar opportunities and are equally able to benefit from any comparative advantage. The ways to analyse this are through the examination of outcomes – can people obtain housing finance on similar terms where similar risks are involved? – rather than looking at whether or not they do this through similar instruments and organisations.

The evidence of the last few years is that the pressures of deregulation have affected housing finance markets throughout the Union. In some cases the impact has been to remove any special circuit and to address the resultant risks through insurance. In others the liberalised market is tending to compete away the benefits associated with the special circuit of housing finance. There are, however, very different pressures in the different countries, arising in particular from the nature of government involvement: Are specific subsidies still being provided? Is the government prepared to guarantee loans? Are tax benefits available for particular types of housing? Where the answers to these questions vary not only will instruments and institutions differ but so also will costs, prices and access.

There is considerable evidence of change. There is also considerable evidence that systems are still in transition – generating further differences as well as convergence. The fundamentals of the legal definitions of property rights and their enforcement show no signs of modification. When and if convergence in this context does occur the reason for it will not be housing related but will be the result of far more general tensions.

Overall the evidence suggests that it is still wider economic and global pressures rather than the development of the Union itself which are the engines of change. What is required now is detailed empirical evidence on the costs and availability of funds across countries as well as on access and profitability among organisations in the finance market. Such evidence is likely to show that there is a long way to go before there is a level playing field in housing finance.

Notes

1 Experts in housing policy and finance were asked to complete questionnaires describing the existing situation in their countries and how this situation was seen to be changing in three different contexts: an overview of changes in financing owner-occupation, for the ENHR in 1995 (Whitehead 1996); a study of the impact of taxation and subsidy on tenure structure, for the Council of Mortgage Lenders (Freeman *et al.* 1996a) and research for the Swedish Housing Commission's study of comparative housing finance (Turner *et al.* 1996).

References

Bartlett, W. and Bramley, G. (eds) (1994) *European Housing Finance: Single Market or Mosaic.* Bristol: School of Advanced Urban Studies.

Basle Committee of Banking Supervisors (1988) *International Convergence of Capital Measurement and Capital Standards.* Basle: BCBS.

Bowen, A. (1994) 'Housing and the Macro-Economy in the United Kingdom', *Housing Policy Debate* 5, 3: 241–251.

British Bankers' Association (1990) *The EC Banking Directives of 1989: A Compendium.* London: BBA.

Commission of the European Communities (1984) *Draft Directive on Freedom of Establishment and Services in the Field of Mortgage Credit,* COM(84)730. Brussels: CEC.

Commission of the European Communities (1989) *Second Banking Coordination,* 89/646/EEC. Brussels: CEC.

Freeman, A., Holmans, A. and Whitehead. C. (1996a) *Is the UK Different? International Comparisons of Tenure Patterns.* London: Council of Mortgage Lenders.

Freeman, A., Holmans, A. and Whitehead, C. (1996b) 'Evaluating the Impact of Tax and Subsidy Frameworks on Housing Tenure Patterns', *European Mortgage Review* 4 (Sept): 18–25.

House of Lords, Select Committee on the European Communities (1985) *A Common Market in Mortgage Credit*, HL177. London: HMSO.

Miles, D. (1994) *Housing, Financial Markets and the Wider Economy*. Chichester: Wiley.

Turner, B., Jakobsson, J. and Whitehead, C. (1996) 'Comparative Housing Finance', in *Bostadspolitik 2000 – fran producktions-till boendepolitik*. Stockholm: Statens Offentliga Utredningar.

Whitehead, C.M.E. (1996) 'Trends in the Provision of Housing Finance in Sixteen Countries', *European Mortgage Review*. 3 (July): 18–22.

3

HOUSING INVESTMENT IN EUROPE

Explanations for differences between countries

Michael Oxley and Jacqueline Smith

In this chapter, we will examine several ways of measuring housing investment and comparing rates of housing investment between countries. We raise questions about the usefulness and reliability of various sets of housing investment data which might be used to make comparisons within the European Union. We discuss theories of housing investment and generate a number of relevant hypotheses. We consider various possible explanations for differences in investment levels, such as: differences in data compilation and the comparability of the statistics; the role of demographic and macroeconomic factors; and policy effects. In relation to demographic and macroeconomic factors, we present results of some econometric modelling. Finally, we provide a critique of the methodology, as well as an appraisal of the nature of alternative explanations, of differences in investment levels.

Housing investment

The ability of a nation to satisfy either housing needs or demands is ultimately related to the physical housing stock of the nation. Increases in the size and quality of this stock amount to physical investment in housing. Low levels of investment may contribute to a series of housing problems such as overcrowding, homelessness and restricted access to good quality dwellings. High levels of investment will boost the capacity of the stock and give the potential for an improvement in housing conditions.

It would be wrong to promote a simple scenario of low housing investment, 'bad'; high housing investment, 'good'. Many other factors do, of course, have to be considered to judge the requirements for, and the consequences of, housing investment. However, the reasons why nations invest at different rates are important to our understanding of the causes of crucial housing problems.

Housing investment can be measured in several ways. Three indicators of housing investment, in particular, have been the subject of collection internationally and are, potentially, a source of information for comparative international analysis. These three measures are:

1   Gross Fixed Capital Formation (GFCF) in residential buildings.
2   Dwellings constructed per 1,000 inhabitants.
3   Net additions to the stock per 1,000 inhabitants.

Each of these indicators will be examined in turn.

Gross Fixed Capital Formation as a percentage of GDP

One measure of housing investment is Gross Fixed Capital Formation (GFCF) in residential buildings as a percentage of Gross Domestic Product. The data source is the United Nations Economic Commission for Europe *Annual Bulletin of Housing and Building Statistics*. In this publication the definition for GFCF in residential buildings is given as follows:

> Value of work put in place on the construction of residential buildings, including major alterations in and additions to such buildings, but excluding the value of land before improvement. Expenditures in respect of the installation of new permanent fixtures are included.
>
> (United Nations 1993: 156)

In supplying data to the United Nations, individual countries make a calculation for investment in the construction of new dwellings, residential buildings and in some cases buildings purchased with the aim of transforming them into dwellings. This includes both public and private investment. The estimates for the investment in new dwellings are calculated excluding the land value of the 'plots' connected to those dwellings. The second major item included is an estimate of expenditures on improvements and major renovations on existing dwellings. This is, more specifically, structural additions and enlargements rather than general maintenance expenditures. Finally, service costs and expenditure related to the production and purchase of the dwellings, such as architects', solicitors' and surveyors' fees, are not included in the GFCF in housing, but are in the total GFCF calculations.

Dwellings constructed per 1,000 inhabitants

This is a measure of dwellings that have been constructed in the course of a year in each country. That is the number of dwellings that have been completed, rather than those authorised, started or under construction, expressed per 1,000 inhabitants. Detailed descriptions of definitions of 'dwellings' and any country-based adjustments are given in the United Nations' *Annual Bulletin*.

This measure is used in addition to the first so that comparisons can be made regarding the actual numbers of dwellings that have been built in each country, as an alternative to the value of new construction and improvements.

Net additions to the stock per 1,000 inhabitants

Dwellings constructed per 1,000 inhabitants gives a gross figure of additions to the stock as a result of new house-building. However, an issue that must be addressed is that some of this building will replace dwellings that have been demolished, and further changes in the stock will occur as a result of changes in use. Therefore, data are also available for a limited number of countries for 'net additions to the stock'. This shows, then, the net changes in the stock taking into account total stock increases (i.e. demolition and negative changes in use). The resulting 'net additions' gives a net total of new stock that is added, per annum.

While these indicators are essentially all measuring different specific components, they can be used broadly as alternative ways of representing housing investment. That is not to say that they are interchangeable measures which may replace each other. Indeed direct comparisons between the indicators cannot be made, because of what is included in the figures and how they are expressed. This is particularly true of the data for GFCF in residential buildings, which includes estimates of improvement work, and the data for net additions which account for decreases as well as increases in the stock.

Relationships between housing investment indicators

The relationship between the housing investment indicators varies from country to country. Housing investment as a percentage of GDP will be influenced by the size of GDP as well as the volume of housing investment. Dwellings built per 1,000 inhabitants will be influenced by population size as well as house-building. The difference between housing investment in total and house-building is largely a function of the volume of resources going into improvement work. Thus, differences in a variety of items may help to explain variations in the degree of association between H (GFCF as a percentage of GDP) and D (dwellings constructed per 1,000 inhabitants) in Table 3.1. In Belgium, France, Greece and Ireland there is a high and significant degree of association between the measures. But in some countries, including the UK, correlation on the basis of data for 1970–1992 is small.

The degree of association between H and N (net additions to the stock per 1,000 inhabitants), also shown in Table 3.1, again varies considerably between countries. The UK stands out in both sets of correlations as a country with very low coefficients.

In explaining differences in housing investment between countries one indicator (H, D or N) will not serve as a good proxy for another given the varying degrees of association displayed by the data in Table 3.1. This suggests that in

Table 3.1 Correlation coefficients of housing investment
indicators

Country	H and D	H and N
Belgium	0.9414*	
Denmark	0.546	0.5565
France	0.8509*	
Germany	0.3405	0.3449
Greece	0.7677*	
Ireland	0.7512*	0.9061
Italy	0.6036	
Netherlands	0.4947	0.4083
Portugal	−0.0519 .	
Spain	0.3385	
UK	0.0272	0.0386

Notes
H = GFCF as a percentage of GDP.
D = Dwellings constructed per 1,000 inhabitants.
N = Net additions to the stock (total stock increases – total stock
    decreases).
* = significant at 1 per cent level.

any statistical testing of hypotheses there will be variations in results according to
which indicator is used as the dependent variable. It also suggests that ultimately
there might be different explanations for housing investment differences on the
basis of different indicators.

Housing investment in the European Union
1970–1992

Gross Fixed Capital Formation in Residential Buildings as a percentage of Gross
Domestic Product was lower in the UK in each year from 1970–1992 than in
most of the other European Union countries. The minor exceptions were that in
some years Spain, Belgium and Ireland had lower housing investment levels
(United Nations 1993).

Dwelling production was also lower in most years in the UK than a majority of
the other countries. There does appear to be a downward trend for most countries, including the UK, but France, the Netherlands, Spain and Greece built
relatively more houses in every year than the UK.

Table 3.2 shows the average figures for (1) GFCF as a percentage of GDP and
(2) dwellings built per 1,000 inhabitants for the period 1970–1992. It can be
seen that the UK ranks eleven out of eleven on the first indicator and nine out of
eleven on the second. There is clear evidence to support the proposition that
housing investment has been relatively low in the UK.

Net additions to the stock have also been relatively low in the UK. Average net
additions to the stock per 1,000 inhabitants over the period 1970–1990 was 3.8

Table 3.2 Housing investment in Europe, averages 1970–1992

Country	Housing investment as % GDP	Rank	Dwellings per 1,000 inhabitants	Rank
Belgium	4.83	9	4.84	8
Denmark	4.95	7	6.46	6
France	6.06	2	7.42	4
Germany	5.93	3	6.32	7
Greece	6.1	1	14.3	1
Ireland	4.95	7	6.94	5
Italy	5.56	4	3.96	11
Netherlands	5.43	5	8.27	2
Portugal	3.95	10	4.55	10
Spain	5.14	6	7.48	3
UK	4.5	11	4.69	9

Sources: United Nations (1993), own calculations.

in the UK compared with 6.5 in Germany and 6.7 in the Netherlands (United Nations 1993).

Explaining differences in housing investment between countries

In broad terms there are two ways of approaching an explanation of the differences revealed in the data. The first is to argue that the statistics do not reliably reflect real differences in outcome, but are rather a result of differences between countries in data collection. The second is to assume that the data does represent real and significant differences between countries and apply theories of the determination of housing investment to the problem.

The first approach implies more than a healthy scepticism about the data. It implies a definite conclusion that the data are either unreliable or internationally incompatible or both. Healthy scepticism is warranted. The data are compiled first for the purposes of national accounts and national recording. United Nations statisticians collate the data and apply common definitions. There is within the UN accounting conventions a systematic attempt to be consistent but we cannot be 100 per cent sure that this consistency has been achieved.

Differences between countries in both the reliability of the data and the interpretation that may reasonably be put on the data might be important. This could, for example, be particularly the case for the value of the improvement work which is a component of GFCF. The interpretation here of 'major additions and alterations' is almost certainly the subject of some national variations. The very recording of this information may vary according to who does the work. Significant 'do-it-yourself' activity would lead to an under-recording of housing investment. The role of the 'black economy' could have important effects on both improvement and house-building work.

The net additions data are potentially the most interesting as they might represent the clearest indication of changes in the capacity of the housing stock. However, conversions and demolitions which are crucial to its analysis are both the subject of varying practices and the data is in any case only available for a relatively small number of countries.

While there may be ample casual empirical evidence for the 'healthy scepticism' approach with respect to the data, there is insufficient evidence to suggest that the incompatibilities are such that the data should be dismissed and that one should reach the conclusion that we have no reliable basis to compare housing investment levels. There may be a good case for a research project which is explicitly a detailed examination of the data compilation, its reliability and transferability. In the absence of such a study, the rest of this chapter rests on the assumption that the data presented do show significant international variations.

Theories of housing investment

While there are several broad theoretical approaches to comparative housing analysis, theories specifically related to explaining differences in housing investment levels are sparse. This paucity of hypotheses is linked to the lack of theories of housing investment generally including attempts to explain intertemporal differences within countries.

Several economic analyses relevant to housing investment adopt an approach which gives a central role to cyclical fluctuations. The effects of long-term building cycles on construction investment have been well documented (see in particular Parry Lewis 1965). Work has been done to test long-term fluctuations in investment and building cycles in the built environment (see Harvey 1978; Butlin 1964; Kuznets 1973; and more recently Ball and Wood 1994). Badcock (1984) argues that these cycles are due to immigration and the rate of household formation. Parry Lewis (1965) also found that the key to long-term fluctuations in construction activity is in 'population, credit and shocks' – where shocks include external factors such as wars, natural disasters, etc. As such it is impossible to forecast investment accurately because of the effects of random events.

However, more generally the rate of economic growth and the growth cycle have been argued to have a significant effect on investment fluctuations. In simple theoretical terms 'a record of rapid growth tends to promote high investment, and a low growth record or falls in output promote cuts in investment' (Black 1982: 74). In practice, however, these cycles are more complicated. In an examination of ten countries by Kuznets (1973), increases were found in investment levels in times of economic growth. He explains that this trend is not that surprising because, 'economic growth meant large increases in per capita product, and one would expect that the national savings rate rose; and, with a greater supply of capital funds, one would expect that the domestic capital formation proportions also rose' (1973: 128). However, in more detailed analysis several exceptions to this general statement were discovered and led Kuznets to conclude that: 'the combination of the broad association between economic growth and the rise in capital formation proportions occurs only with significant disparities in timing of the two variables' (1973: 129).

Much conflicting evidence exists in relation to investment and growth rates. Indeed, the introduction to a study on building investment and economic growth by Ball and Wood (1994: 1) cites studies (Aschauer 1989; De Long and Summers 1991) which respectively argue that there is a highly positive, and negative relationship between the two factors. To facilitate an independent analysis of housing investment for this study, economic growth will be used as a variable in some of the econometric testing.

In relation to building cycles it might also be useful to consider the effects of population and household formation, introduced above. Demographics and, in particular, population growth is argued to be a specific determinant of increases in investment (Kuznets 1973; Black 1982; Hillebrandt 1974). This is because, 'the higher the rate of increase in population and labour force, the greater requirement for material capital to equip additional workers' (Kuznets 1973: 10). More specifically, investment in housing will be influenced by demographic changes due to changes in the birth rate, the death rate or rate of household formation.

An increase in the rate of household formation requires an increase in housing investment if 'housing need' and 'housing demand' are to be met. One study by Holmans (1995) assessed future levels of need and demand (and therefore likely levels of future housing investment) using demographic analysis. Future levels of new dwellings needed were estimated using a 'modified net stock' method. This method takes into account effective demand for houses for private owners and needs for social housing; the latter is based on both existing unmet needs and newly arising need (largely a function of increases in the number of households and changes in the types of households). This methodology assumes, significantly, that the number and composition of households in a country is a major variable in determining future need and therefore future housing investment. While this study is not concerned with forecasting, Holmans' methods exemplify the important link between household growth, need and housing investment.

Studies examining the effects of population change on housing investment, however, have not had such significant results as those looking at household growth. For example, Ball and Wood's study stated that trends in housing investment weighted by population did 'not seem to be greatly affected . . . suggesting that national level population changes are not the prime cause of variations in housing investment' (1994: 6). They do not dismiss the relevance of population effects on housing investment but claim that its effect, 'is transmitted through several inter-linked processes which weaken the direct relationship' (1994: 6). In practice the rate of household formation is much more likely to have a positive effect on housing investment, as illustrated in Holmans' methodology, than is population growth. To clarify the effects of household and population growth in this study, demographic factors will also be included in the later statistical analysis.

There are a number of other basic economic principles that are often used to explain trends in investment. Many of these are more frequently used to explain investment in plant and machinery, and especially the manufacturing sector, rather than housing investment. Much of the existing empirical work therefore reflects this bias. There are relatively few works that have theorised on the economics of housing investment. While the economic theories discussed below are more frequently related to manufacturing investment, it will be useful briefly to examine their principles to determine possible relationships with housing investment.

The accelerator theory is commonly used in economics to explain changes in investment. The theory states that investment will be related to the rate of change of output. If the demand for any consumption good increases, then the demand for investment goods used in its production will increase at a greater rate. Therefore, future investment is calculated by multiplying the increase in output, over a fixed time, by a fixed capital/output ratio (Griffiths and Wall 1993). A number of criticisms have been levelled at this theory. Many of the assumptions of the theory are unlikely to hold in practice. For example, the assumption that no excess capacity exists, that the capital/output ratio is constant and that future expectations of industry do not play a significant role. Therefore, even in relation to non-housing investment the accelerator theory has limitations.

The theory has been largely rejected in respect to housing investment; Black describes the theory as a 'quite useless' explanation of investment as a whole and though it has some use in application to stocks and work in progress he states that it cannot be applied to investment in dwellings which, 'has a cycle of its own' (1982: 77). Hillebrandt (1974) used the accelerator principle to explain investment in industrial and commercial building, showing it to be a useful application provided increases in demand were anticipated (1974: 63 for full explanation). Hillebrandt states that the accelerator principle is less useful when applied to housing investment because changes in demand for housing are more difficult to predict, and housing demand is very dependent on government policy.

Partly to overcome the problems with the assumptions of the accelerator described above, the capital stock adjustment model was developed to explain investment more accurately. This model states that investment is related positively to changes in levels of output and negatively to the existing capital stock. Therefore, investment for the next period will increase in relation to the level of past output, but will be reduced proportionally by the volume of capital stock already existing. This model has proved to be considerably more useful in application to manufacturing fixed investment, and in particular when capacity utilisation variables were introduced (see Kennedy 1986; Ford and Poret 1990).

Studies do not appear to have been carried out specifically on housing investment. However, the adjustments to the accelerator principle make this model much more relevant to housing. Although problems of predicting changes in the demand (or need) for housing remain, indicators of previous construction and existing size of dwelling stocks could provide a clearer analysis. Further analysis of the effects of the size of existing stocks on investment in housing will, therefore, be included in the later statistical testing.

A further economic factor that may affect investment is the rate of interest. An investment decision is based on whether the rate of interest on borrowing for investment will be more than covered by the annual return on the initial capital outlay. If the interest rate is lower than the return on capital (or 'marginal efficiency of investment'), the investment is profitable and therefore worthwhile. A large amount of emphasis in this theory is placed on expectations. As expectations are volatile, at any interest rate level there will be changing expectations of future returns. Griffiths and Wall note that consequently 'it may be via expectations that interest rates exert their major influence on investment', but that this possibility reduces the 'closeness of any statistical fit between the interest rate and investment' (1993: 296).

Studies examining this relationship have difficulties selecting an appropriate interest rate. Many UK studies undertaken have not been able to illustrate a close relationship, indicating that investment is actually, 'interest-inelastic' (Griffiths and Wall 1993: 296). Griffiths and Wall note that from a number of studies on fixed investment (and particularly manufacturing investment) those introducing lag structures were the most effective. Other studies also only reported a weak negative relationship (e.g. Turner 1989).

Bank of England evidence (Easton 1990), however, has shown that interest rates have had some considerable effect on investment, particularly housing. It reports that the restructuring of the housing market due to deregulation in the mortgage market means that housing takes up a much greater role in the personal sector balance sheet, resulting in a greater interest rate effect on investment. A Bank of England model simulation of the UK economy illustrates the impact of an increase in interest rates on investment and residential investment. Their model shows that after a 1 per cent point increase in all interest rates, investment would fall by 2.2 per cent over two years and by 2.8 per cent over three years. In addition, private residential investment would fall more substantially over the periods, by 3.1 per cent and 4.1 per cent respectively. The author notes, however, that all interactions are subject to uncertainty due to the effects of expectations, but results can provide some qualification of the interest rate effects on investment (Easton 1990: 203).

Other economic principles that will both affect gross investment and housing investment include profitability, the degree of uncertainty in the sector and the economy as a whole. Finally, the management of the economy through public policy has been shown to have an effect on investment. The stop-go nature of public and economic policies causes uncertainty in the economy which has a negative impact on confidence and can discourage investment.

There is evidence also to show that, in the past, government investment in housing was undertaken to regulate construction in the private sector (Badcock 1984: 139; Hillebrandt 1974: 19–20). In addition there are arguments that relate changes in investment and housing policy to external factors – or 'shocks'.

One of the most thoroughgoing attempts to test economic theories of housing investment on an international comparative basis was undertaken by Burns and Grebler (1977). Using cross-section data for 1963–1970 for thirty-nine countries they used multiple regression analysis to find an explanation for average housing investment as a proportion of average annual Gross Domestic Product. The key explanatory variables were GDP per head, population growth and the level of urbanisation. Their study produced statistically significant results suggesting that the three variables together accounted for over half of the variance in their measure of housing investment (1977: 34). However, they acknowledged that several potentially important variables were excluded from their analysis such as the magnitude of government assistance to housing and the size and age (or condition) of the housing stock.

There has thus been more work on building investment generally than on housing investment. An adaptation of the capital stock adjustment principle to housing, with demographic factors being used as proxies for demand and the housing stock used as an indicator of 'current capacity', may offer some promise for explanations of housing investment. The role of economic and demographic variables has been demonstrated by Burns and Grebler. Systematic testing of hypotheses which combine economic, demographic and housing stock variables seems likely to offer a more complete explanation.

Hypothesis for statistical testing

In the sections below, an examination of several hypotheses related to an economic and demographic analysis of investment is presented and tested for several countries in the European Union for the period 1970–1992. The section above outlined certain theories about economic determinants of housing investment and a number of different factors were identified as possible determinants. The macroeconomic and demographic factors to be tested in this chapter are those that have been identified as meeting two criteria: first, that theory suggests the factor might have a significant effect on the dependent variable and, second, that the nature of the factor lends itself to statistical testing. Omitting some factors is not an indication that their potential effects on housing investment outcomes are ignored; rather that it is not always possible to analyse all factors using the same methodology.

The hypothesis testing has taken both a cross-section and time series approach. The former uses averaged data over the period 1970–1992 for each country to make comparisons between countries. The time series analysis considers variations over time within countries.

Four main factors have been indicated as possible housing investment determinants that might be tested statistically: total investment, economic growth, demographic growth and the size of the dwelling stock in the different countries. Four hypotheses which examine the relevance of each of these factors to explanations for variations in housing investment are tested. In addition to this, a number of the variables outlined above will be tested together to examine their combined effect on housing investment using multivariate analysis. Multivariate models play an important role in combining the various hypotheses to examine the effects of explanatory factors together on housing investment. Two multivariate models are examined.

First, a 'need' model will examine the combined effect of demographic changes and the size of the stock on housing investment. This model tests the statistical relationship between both demographic growth and the size of the stock, and housing investment. Combining demographic factors with the size of the dwelling stock indicator can test the effect on housing investment of changes in the number of households and the population in need of a dwelling, with the number of dwellings available in the housing stock. By combining the hypothesis relating to the dwelling stock and demographic growth, this model can examine the effect on investment levels of the size of the stock and levels of need as indicated by household and/or population growth. This model, therefore, goes some way to testing the theory of the 'capital stock adjustment model' discussed earlier.

Second, a 'need and growth' model will examine the combined effect of demographic change variables, the size of the stock and economic growth on housing investment. By adding a variable that measures economic growth to the need model, the statistical testing can also evaluate how growth in an economy, together with households' need for dwellings and the number of available dwellings, affect housing investment. The 'need and growth' model therefore assumes that an extra stimulus and an improved potential for meeting need come within a housing system when higher growth results in more resources.

Housing investment and total investment

It might be argued that housing investment is simply a reflection of investment generally in an economy and that there is a high degree of association between housing investment and other forms of investment such as manufacturing investment. If this is correct, an analysis of differences in housing investment might become largely an investigation of why investment generally varies from country to country. The means to increase housing investment might also be largely those required to increase investment generally. The emphasis of research would shift from a 'housing study' to an 'investment study'.

There is a well-developed body of literature on the determinants of investment generally and this would then provide the theoretical foundations for the study. Given that housing investment is, *ipso facto*, part of overall investment, it would be extremely surprising if there was no connection between the two. The issue is, how close is the connection?

The hypothesis in question can be formally expressed as:

$H = f(I)$, where
H = Gross Fixed Capital Formation in housing as a percentage of Gross Domestic Product, and
I = Total Gross Fixed Capital Formation as a percentage of Gross Domestic Product.

The cross-section evidence summarised in Figure 3.1 plots the average level of H against the average level of I (during the period 1970–1992) for eleven countries for which adequate data was available.

Clearly, there is no simple linear relationship between the two variables. The R^2 figure suggests that less than 14 per cent of the variation in housing investment *between* countries can be explained by overall variations in GFCF. The UK is

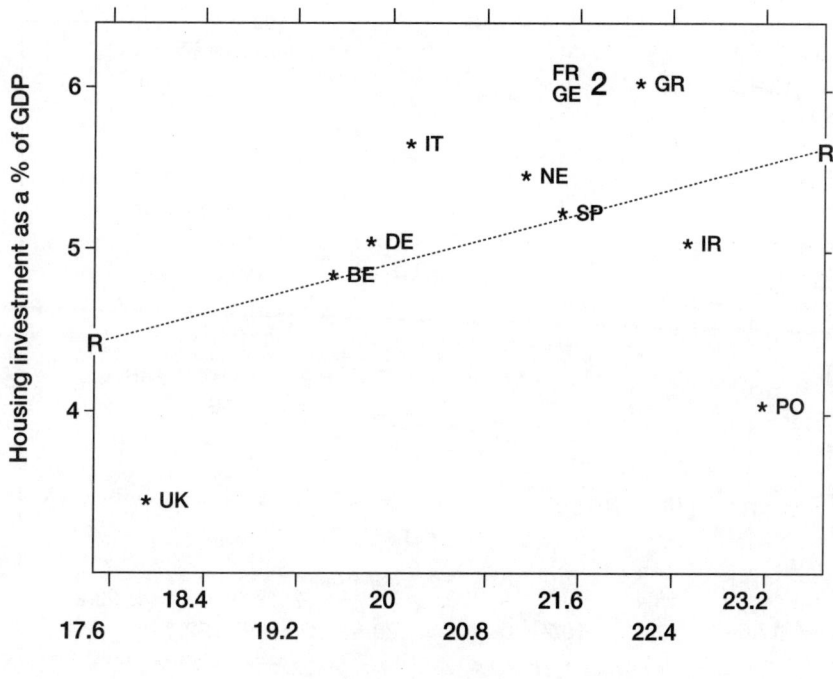

Total investment as a % of GDP

Figure 3.1 Housing investment[a] and total investment[b]

Notes
a   Per cent of GDP devoted to Gross Fixed Capital Formation in housing.
b   Gross Fixed Capital Formation.
    Averaged data 1970–1992.
    For key to notation see Appendix p. 56.

distinguished by low housing investment and low investment generally but Portugal, for example, has relatively low housing investment and the highest level of overall investment. If we confine our interest to dwellings completed we may express the hypothesis that:

D = f (I), where
D = Dwellings completed per 1,000 population.

The relationship between D and I is shown in Figure 3.2. Again both D and I are averages for each country for the period 1970–1992. The R² statistic suggests that again approximately 14 per cent of the variations *between* countries in dwelling production is associated with variations in total GFCF.

In addition, we can investigate the (time series) relationships between H and I, and between D and I *within* countries. In Table 3.3, we present the results of

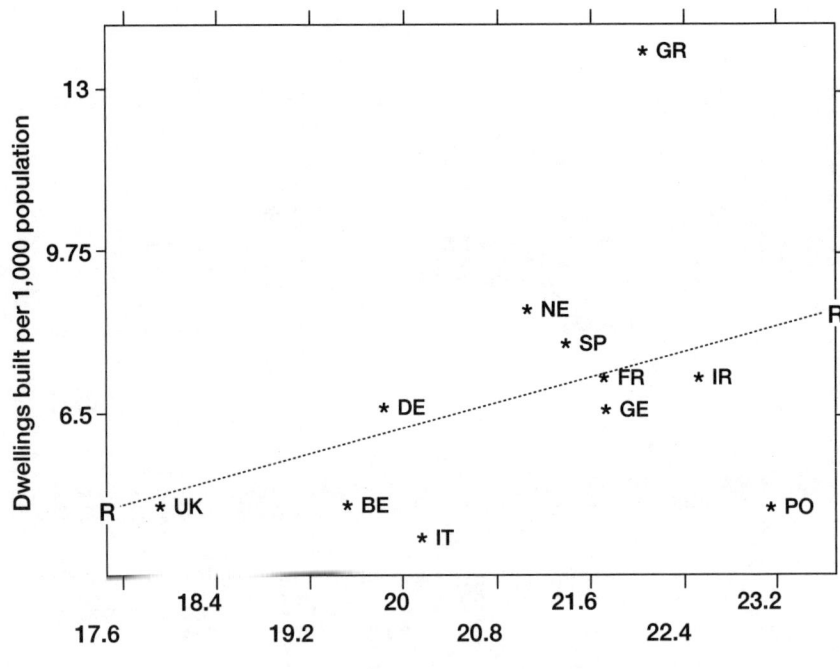

Figure 3.2 Dwellings built[a] and total investment[b]

Notes
a   Dwellings completed per 1,000 population.
b   Gross Fixed Capital Formation.
     Averaged data 1970–1992.
     For key to notation see Appendix p. 56.

Table 3.3 Housing investment and total investment: summary of regression analysis for the hypotheses that $D = a_1 + b_1I$ and $H = a_2 + b_2I$

	D			H		
	R^2	F	b_1	R^2	F	b_2
UK	0.30	*	0.47	0.48	*	0.19
Germany	0.54	*	0.76	0.059		0.07
Netherlands	0.18	*	0.36	0.017	*	0.11
France	0.66	*	0.72	0.69	*	0.41
Belgium	0.70	*	0.59	0.67	*	0.45
Italy	0.04		0.26	0.54	*	0.31
Portugal	0.26		0.06	0.45	*	0.39
Spain	0.09		0.33	0.07		0.14
Greece	0.50	*	1.06	0.56	*	0.42
Ireland	0.35	*	0.21	0.78	*	0.26
Denmark	0.75	*	0.62	0.65	*	0.35

Note
* Significant at 0.05.

linear regression analysis on a country by country basis for the period 1970–1992. For each country equations were generated for:

$H = a_1 + b_1I$, and
$D = a_2 + b_2I$, where
a_1 and a_2 are constant terms and b_1 and b_2 are coefficients for I.

For most countries there is a significant statistical relationship between variables on the basis of an F test for the significance of the equations at a 0.05 confidence level. The strength of the relationships is reflected in the adjusted R^2 data which suggest high levels of association over time between investment overall and housing investment for several countries. The coefficients for the independent variable vary considerably between countries.

Therefore, more GFCF does mean more housing investment in all countries. However, the strength of the relationship varies a lot between countries and GFCF clearly does not explain all the variation in housing investment. We can thus conclude that housing investment differences cannot be explained simply by differences in total investment.

Housing investment and the growth of Gross Domestic Product

Higher growth of output in an economy might be expected to generate additional demand and higher real incomes and thus boost housing demand and generate additional resources from which the demand might be met. Economic growth might thus be expected to result in additions to housing investment. Thus, we might hypothesise that:

H = f (g), and
D = f (g), where
g = annual growth of Gross Domestic Product (GDP)

Data by country for H and g is shown in Figure 3.3 and for D and g in Figure 3.4. The economic growth data have been taken from the OECD's *Economic Outlook* (OECD 1993).

Analysis of the data in Figures 3.3 and 3.4 shows that there is, statistically, very little relationship between annual growth of GDP and either H or D. The relevant R^2 statistics are both very low reflecting the very low (and negative!) correlations.

Regression equations were run on a country by country basis to investigate the relationships between g and each of H and D. For those countries where data

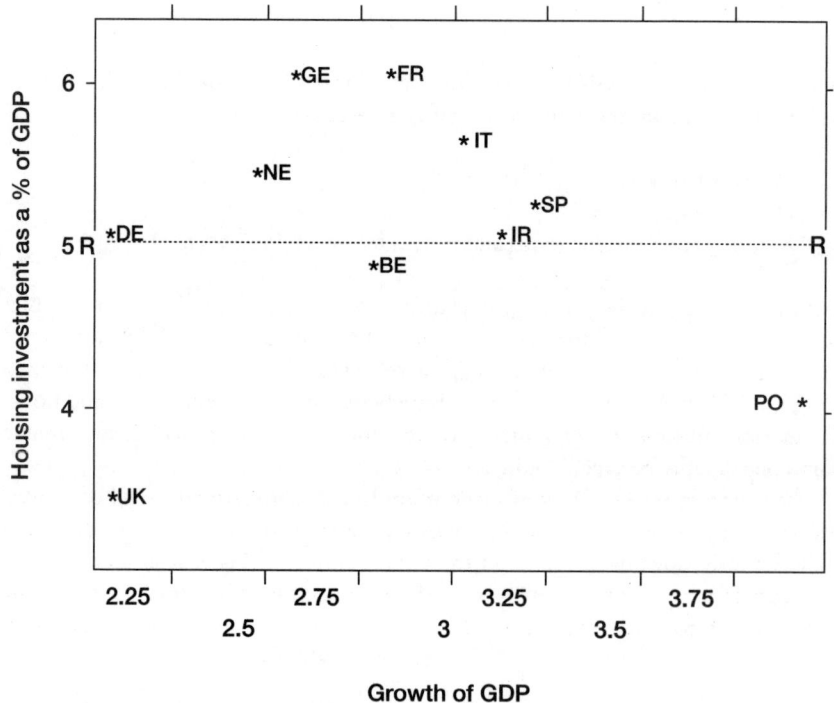

Figure 3.3 Housing investment[a] and growth of GDP

Notes
a   Per cent of GDP devoted to Gross Fixed Capital Formation in housing.
    Averaged data 1970–1992.
    For key to notation see Appendix p. 56.

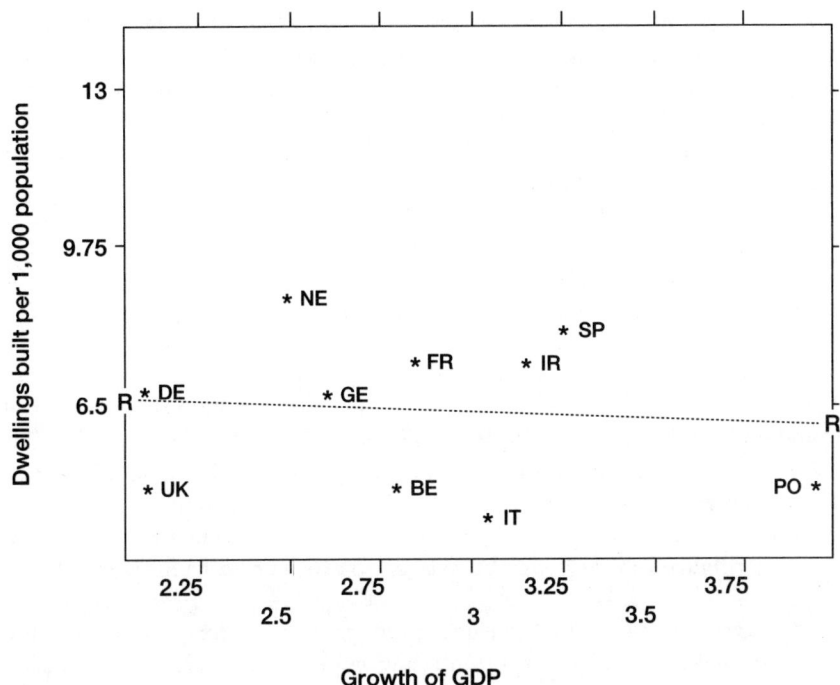

Figure 3.4 Dwellings built[a] and growth of GDP

Notes
a   Dwellings completed per 1,000 population.
     Averaged data 1970–1992.
     For key to notation see Appendix p. 56.

were available, net additions were also investigated as the dependent variable. No statistically significant relationships were found.

The lack of such a relationship is associated with some contrasting national experiences. Denmark with a similar growth rate to the UK (and a lower growth of GDP per head) has undertaken significantly more housing investment while the high growth rate in Portugal has been associated with a low level of housing investment.

Differences in growth rates between countries do not appear to explain differences in housing investment and within countries higher rates of growth do not result, in any systematic fashion, in higher levels of housing investment.

Housing investment and demographic factors

Demographic data have been taken from a number of sources which include OECD National Accounts (OECD 1992), Eurostat (1992), European Commission (1993) and national governments.

Higher levels of demographic growth in a country might be expected to result in more investment. This could be because of the effect of demographics on demand or because of the effect on need and the governmental response to this. There might be a time lag involved. The appropriate lag is difficult to determine and might vary between countries. If demographic change is forecast reasonably well, the lag might be fairly short.

The following hypothesis was tested:

$H = f(n+)$, where
$n+$ = the proportionate change in population over the previous four years.

The correlation coefficient was found to be only 0.06 for the averaged values of H and $n+$. Clearly there is very little relationship between the variables. With D as the dependent variable, i.e. $D = f(n+)$, the relationship was found to be stronger. Taking the average values of D and $n+$ for the period considered, there is a correlation coefficient of 0.41 and thus about 17 per cent of the variation between countries in dwellings built is associated with differences in rates of population change.

The stronger effects for dwelling production than for investment raise questions about the nature of the housing improvement part of the housing investment indicator. It might be that this component is not very responsive to demographic factors whereas the house-building component is more responsive. However, this hypothesis was not tested.

Numbers of people in the population may not be the most appropriate indicator of demographic change in housing analysis. A more appropriate measure, of course, might be the change in the number of households. The effect of household change was investigated for those countries for which appropriate data was available.

The following hypothesis was tested:

$H = f(n)$, where
n = the proportionate change in the number of households over the previous four years.

In Figure 3.5, the results of examining the average values of H and n are shown. The R^2 value suggests that approximately 14 per cent of the change in housing investment is associated with household change.

The following proposition was also considered:

$D = f(n)$

The averaged values of D and n shown in Figure 3.6 suggest a significant association between household change and numbers of dwellings built. The R^2 figure is

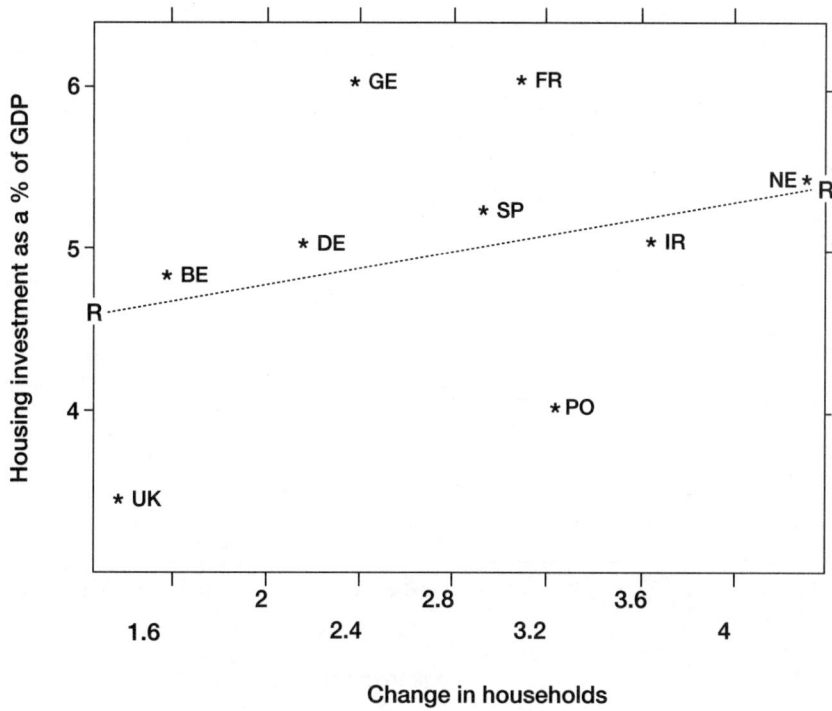

Figure 3.5 Housing investment[a] and change in households

Notes
a   Per cent of GDP devoted to Gross Fixed Capital Formation in housing.
    Averaged data 1970–1992.
    For key to notation see Appendix p. 56.

0.47, that is, almost half of the variation in dwelling production is associated with changes in the number of households.

We also investigated, within countries, relationships on a time series basis between measures of housing investment and measures of demographic change. We tested a variety of models using D, H and N (net additions to the stock per 1,000 population) as the dependent variables and measures of change in the total population and numbers of households as independent variables. Tests involving lagged effects of demographic change were also conducted. The results varied considerably both from country to country and according to which variables were used.

The mixed and contrasting results from the demographic analysis on a country by country basis might genuinely reveal different processes at work in relating demographic change to housing investment, but it is more likely that one needs to combine demographics with other factors in more complex models to pick up its significance.

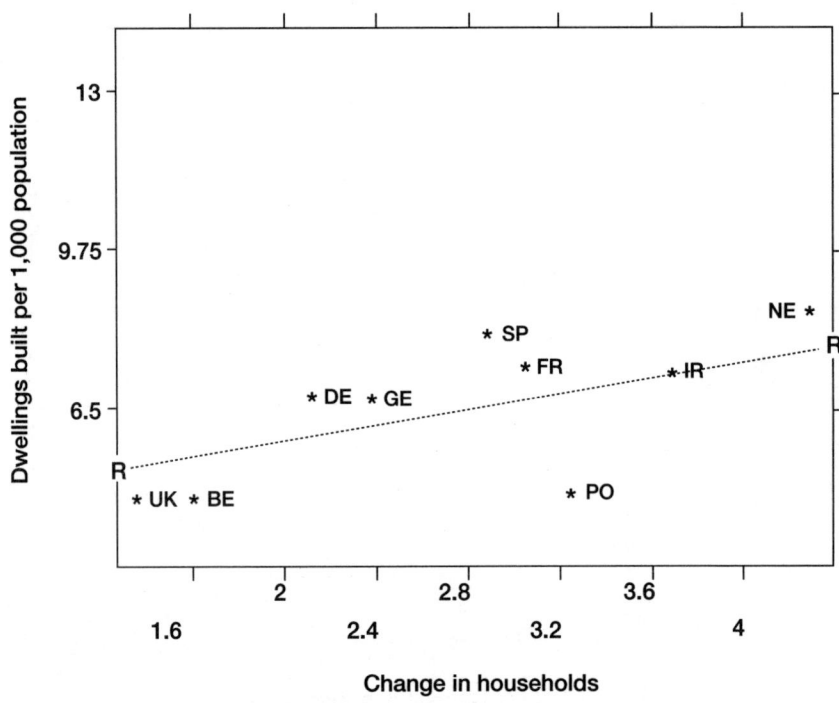

Figure 3.6 Dwellings built[a] and change in households

Notes
a   Dwellings completed per 1,000 population.
    Averaged data 1970–1992.
    For key to notation see Appendix p. 56.

Housing investment and the size of the housing stock

In a country which has a comparatively large housing stock relative to its population, it might be assumed that the need for extra housing investment is low compared with a country which has a small housing stock. Thus we might postulate that:

$H = f(S)$, where
S = the number of dwellings in the stock per 1,000 population.

We can hypothesise that H is a negative function of S so that as the stock gets bigger additional investment falls. Results reported in Figure 3.7, however, suggest a small positive correlation between averaged H and S, with a correlation coefficient of 0.17 and R^2 of 0.03. This might be a consequence of larger housing stocks inviting higher levels of improvement work.

On the basis that the need for additional dwellings might fall as the dwelling

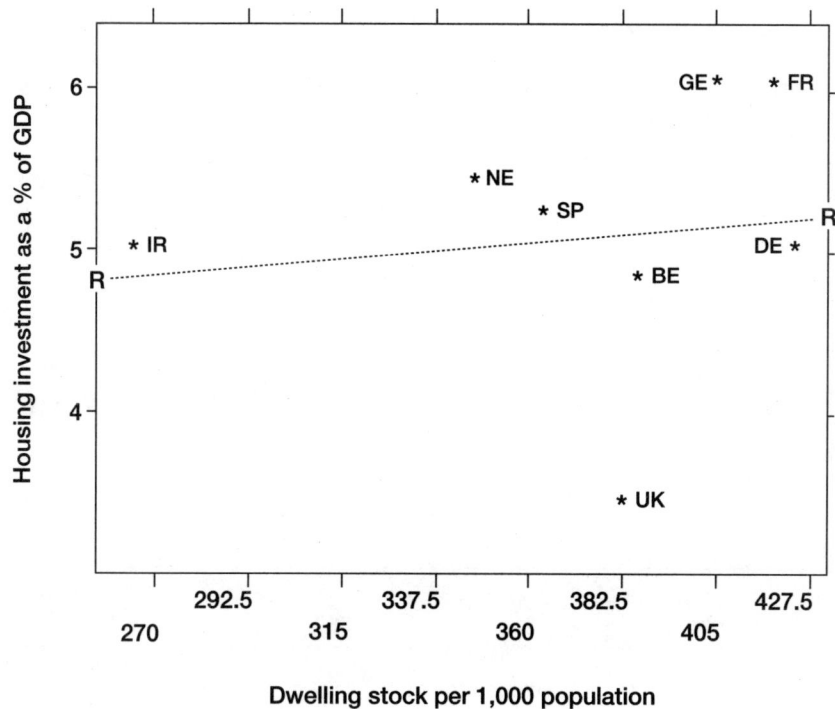

Figure 3.7 Housing investment[a] and dwelling stock

Notes
a   Per cent of GDP devoted to Gross Fixed Capital Formation in housing.
    Averaged data 1970–1992.
    For key to notation see Appendix p. 56.

stock becomes larger relative to the population in a country, we might also assume a negative relationship between D and S. The results in Figure 3.8 do suggest a small negative relationship. Here the correlation coefficient is –0.25 and the R^2 0.06. However, as the R^2 values associated with the relationships in Figures 3.7 and 3.8 are so small, little significance should be attached to either set of results. The appropriate conclusion is that neither housing investment nor dwelling production can be explained simply by reference to the size of the housing stock.

A 'need' model of housing investment

Thus far only pairs of relationships have been considered in isolation. A 'need' model of housing investment, as discussed previously, might postulate a positive relationship between investment and demographic change coupled with a negative relationship with the size of the housing stock. Thus we might suggest that:

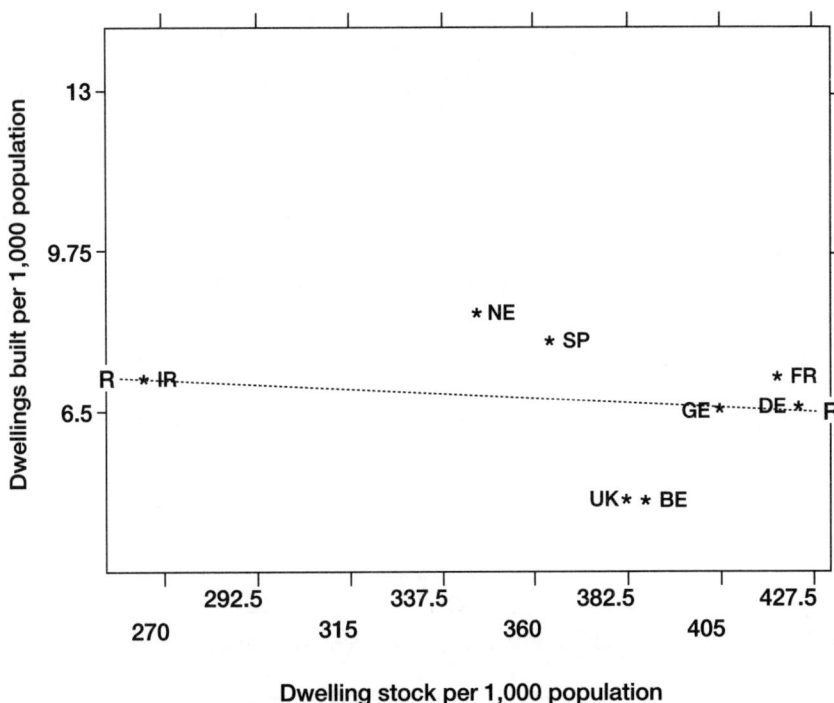

Dwelling stock per 1,000 population

Figure 3.8 Dwellings built[a] and dwelling stock

Notes
a   Dwellings completed per 1,000 population.
    Averaged data 1970–1992.
    For key to notation see Appendix p. 56.

$$H = b_0 + b_1 n + b_2 S$$

We would expect the coefficient for the size of the stock to be negative and all other signs to be positive. However, testing this hypothesis showed the sign for b_2 to be in fact positive. The adjusted R^2 for the equation as a whole was 0.92. The results suggest a strong positive effect from household change. Any 'apparent' positive influence from the stock is very small.

The 'need hypothesis' above was tested on a country by country basis with n and S equal to annual figures for each country. Thus a time series model for each country was applied.

The results with H as the dependent variable with no lags were inconclusive. The model was also applied with two- and four-year lags for the effect of household change. For some countries results were produced with the signs of the coefficients as expected and R^2 as high as 0.7, but for other countries the results were not as expected with an 'incorrect' sign and low R^2.

The results with D as the dependent variable were more consistent. The estimated equations for the UK, France, Germany and the Netherlands are shown in Table 3.4. The effect of a time lag for household change was investigated. The estimated equations both with and without a two-year lag are shown.

Without a lag, the sign for the household change coefficient is not as expected in the case of France. With a lag, the coefficients are all of the expected sign and on the basis of a F test at 0.05 all the equations support the hypothesis. (The results for Denmark and Spain also support the hypothesis; results for other countries include estimates with an 'unexpected' sign for the household change coefficient.) The 'need' hypothesis with a two-year lag for the effect of household change thus works fairly well for several countries and offers some potential for further investigation.

A 'need and growth' model of housing investment

Economic growth might be added to the need model. We can thus combine the effects of demographic change, the size of the stock and economic growth to postulate that:

$$H = b_0 + b_1\, n + b_2\, S + b_3\, g$$

Table 3.4 Dwelling production and housing 'need': summary of regression analysis for the hypothesis that $D = b_0 + b_1 n + b_2 S$

	b_0	b_1	b_2	R^2	*F statistic*
With no lag					
UK	28.24	0.26	−0.63	0.70	20.8
		(1.86)	(−5.37)		
France	28.34	−0.006	−0.049	0.82	41.8
		(0.04)	(−8.74)		
Germany	27.70	0.197	−0.055	0.53	10.2
		(0.921)	(−4.50)		
Netherlands	8.0	0.372	−0.009	0.36	5.8
		(1.98)	(0.084)		
With two-year lag for 'n'					
UK	26.70	0.29	−0.058	0.67	18.0
		(1.31)	(−4.75)		
France	27.86	0.097	0.049	0.83	42.5
		(0.469)	(−9.06)		
Germany	26.94	0.23	0.052	0.52	9.8
		(0.726)	(−4.22)		
Netherlands	15.1	0.118	−0.022	0.20	4.1
		(0.323)	(−1.83)		

Note
t statistics in parentheses.

where b_0 is a constant term and b_1, b_2 and b_3 are coefficients for, respectively, household change, the size of the housing stock and growth of Gross Domestic Product. We would expect the coefficient for the size of the stock to be negative and all other signs to be positive.

The result of testing this at an aggregate level using, for each country, averaged values for n, S and g gave an adjusted R^2 of 0.61. However, as in the 'need' model, the sign for the stock coefficient was positive.

Replacing H by D we can also consider:

$$D = b_0 + b_1\, n + b_2\, S + b_3\, g$$

Again, we would expect the coefficient for the size of the stock to be negative and all other signs to be positive. A cross-sectional analysis applying multiple linear regression to aggregated averaged data gave an equation with an adjusted R^2 of 0.91 but again the sign for the stock was positive. The equation generated was:

$$D = \underset{(7.9)}{1.86} + \underset{(2.9)}{1.37n} + \underset{(0.89)}{0.1S} + 0.35g \; ; F = 25.8$$

T statistics are in parentheses. The positive, although small, effect of the stock seems difficult to explain. With this caveat, we have a fairly comprehensive model which, excepting the incorrect sign for the stock coefficient, explains more than 90 per cent of the variation in housing investment.

The 'need and growth' model was applied to individual countries using time series data. The model was applied both with no time lags and with a two-year time lag for household change. In these results, the signs for the stock coefficients are negative as expected. However, the signs for some of the other coefficients are not as expected. Without any lags, the predicted equations for the UK and Germany have coefficients which each have the expected signs and we have large R^2 and F statistics which are significant at 0.05. With a two-year lag for household change, the equations for the UK and France give expected and statistically significant results.

Tests of the 'need and growth' model with housing investment as a percentage of GDP as the dependent variable were conducted on a country by country basis. Only in the case of the UK did the predicted equation have coefficients with all the signs as expected and then the resulting R^2 and F values were very low. This model is, therefore, less successful in explaining change in investment over time.

By combining demographic factors, the size of the stock and economic growth in various versions of a 'need and growth' model, we are unable to offer a high degree of explanation for differences between countries in housing investment unless a perverse positive effect of the housing stock is accepted. A 'need and growth' model gives satisfactory results for some countries in explaining differences in investment over time but unsatisfactory results for others.

Policy effects on housing investment

Government policy inevitably affects housing investment. It might be assumed that a government or the form of that government's housing policies, influences levels of housing investment according to its degree of involvement in housing or the degree of priority given to housing. Under these circumstances a country where the government has greater involvement in housing or gives it greater priority may have a higher level of housing investment. However, to test this statistically requires finding a variable that is able to quantify levels of government activity in housing. This is clearly very difficult to obtain. Policy effects are not modelled in this work for a number of reasons.

Government expenditure on housing as Burns and Grebler (1977: 36) implied might be a significant variable, but there is a lack of consistent and comparable data on this item. Internationally, accounting conventions vary widely and what is recorded as government or public expenditure on housing in one country will not be recorded in another. What exactly should be included is highly problematic. Some government decisions such as those on interest rates might be picked up by using interest rates explicitly as a variable in a model. However, to assign changes in interest rates purely to policy decisions is problematic given the whole range of national and international influences. Infrastructure costs might be included and are met in varying degrees from country to country from public funds. How should expenditure on housing allowances be recorded? Is this housing expenditure or more general social security expenditure? The problems are immense.

Equally difficult to resolve would be the issues surrounding the measurement of subsidies. If one wished to include the value of housing subsidies as a variable in a deterministic model, the issue of what is and is not a housing subsidy would have to be resolved. If we define a housing subsidy as a government-induced flow of funds which reduces housing costs below what they otherwise would be, the problem of defining the benchmark has to be faced. Some of the most important interventions by government involve 'hidden subsidies' via, for example, land market interventions and loan guarantees which reduce housing development costs but do not involve direct public expenditure (see Oxley and Smith 1996: 40–41, 43–64; Haffner 1994). There are no straightforward ways of resolving these problems.

Policy effects, while significant, perhaps even the most important factors in explaining investment differences, would need to be evaluated by non-quantitative more subjective means. (For a full discussion of the effects of policy on housing investment see Smith 1996.)

Conclusions

It is possible to compare housing investment levels between countries using a variety of indicators. Statistical hypothesis testing reveals a number of plausible and interesting results. The rate of household formation, in particular, is seen to have significant effects on dwelling production. Whether there is, additionally, a reverse causal effect with the availability of dwellings influencing household formation is not so clear. The role of economic growth is, perhaps, unexpected. High rates of economic growth can combine with either high or low rates of housing investment.

The statistical work reported here is relatively simple. Further research should involve non-linear relationships and more complex lag structures. Additional variables could also be incorporated into the models.

The role of government in influencing housing investment could be investigated further by tackling the difficult issues of quantifying public expenditure on housing and housing subsidies on a consistent basis. This would require much more complex modelling work of national housing systems. Alternatively, more intuitive approaches could be adopted by, for example, assigning policy implications to the residuals in econometric models or by using dummy variables to investigate what appear to be contrasting approaches between countries.

While such further work could proceed on the assumption that the United Nations data on Gross Fixed Capital Formation in housing and dwelling production does display meaningful differences between countries, there is additionally a good case for more work on investigating the compilation and comparability of this data.

Appendix

Summary of notation

I   = Gross Fixed Capital Formation (GFCF) as a percentage of Gross Domestic Product (GDP).

H   = GFCF in housing as a percentage of GDP.

D   = Dwellings built per 1,000 population.

N   = Net additions to the housing stock per 1,000 population.

g   = Annual percentage growth of GDP.

n+ = Proportionate change in population over the previous four years.

ıı   – Proportionate change in the number of households over the previous four years.

S   = Number of dwellings in the stock per 1,000 population.

BE = Belgium

DE = Denmark

FR = France

GE = Germany

GR = Greece

IR = Ireland

IT = Italy

NE = Netherlands
PO = Portugal
SP = Spain

References

Aschauer, D. (1989) 'Is Public Expenditure Productive?' *Journal of Monetary Economics* 23: 177–200.

Badcock, B. (1984) *Unfairly Structured Cities*. Oxford: Blackwell.

Ball, M. and Wood, A. (1994) 'Does Building Investment Affect Economic Growth? Some Long-run Evidence from the UK', Discussion Paper 8/94, Birkbeck College, University of London.

Black, J. (1982) *The Economics of Modern Britain: An Introduction to Macroeconomics*. Oxford: Robertson.

Burns, L.S. and Grebler, L. (1977) *The Housing of Nations*. London: Macmillan.

Butlin, N. (1964) *Investment in Australian Economic Development 1861–1900*. Cambridge: Cambridge University Press.

De Long, J.B. and Summers, L.H. (1991) 'Equipment Investment and Economic Growth', *Quarterly Journal of Economics*, 106: 445–502.

Easton, W.W. (1990) 'The Interest Rate Mechanism in the UK and Overseas', *Bank of England Quarterly Bulletin*, 30 (2 May).

European Commission (1993) *Housing in Europe: Statistics on Housing in the European Community*. Brussels: EC.

Eurostat (1992) *Basic Statistics of the Community*. Brussels: OOPEC.

Ford, R. and Poret, P. (1990) 'Business Investment in the OECD Economies: Recent Performance and Some Implications for Policy', *OECD Working Paper 88*. Paris: OECD.

Griffiths, A. and Wall, S. (1993) *Applied Economics: An Introductory Course*, 5th edn, London: Longman.

Haffner, M.E.A. (1994) 'Effects of Financial Policy on Housing Expenses for Owner-Occupiers in Newly built Dwellings: A Comparison of Six Countries in North-western Europe', *Housing Studies* 9, 1: 125–141.

Harvey, D.W. (1978) 'The Urban Process under Capitalism: A Framework for Analysis', *International Journal of Urban and Regional Research* 2 (1): 101–131.

Hillebrandt, P. (1974) *Economic Theory and the Construction Industry*. London: Macmillan.

Holmans, A. (1995) 'Housing Demand and Need in England 1991–2011', *Joseph Rowntree Findings No.157*. York: Joseph Rowntree Foundation.

Kennedy, M. (1986) 'The Economy as a Whole', in A.R. Prest (ed.), *The UK Economy: A Manual of Applied Economics*. London: Wiedenfeld and Nicolson.

Kuznets, S. (1973) *Population, Capital and Growth: Selected Essays*. London: Heinemann.

OECD (1992) *National Accounts*. Paris: OECD.

OECD (1993) *Economic Outlook*. Paris: OECD.

Oxley, M. and Smith, J. (1996) *Housing Policy and Rented Housing in Europe*. London: E. and F.N. Spon.

Parry Lewis, J. (1965) *Building Cycles and Britain's Growth*. London: Macmillan.

Smith, J. (1996) 'What Determines Housing Investment? An Investigation into the Social, Economic and Political Determinants of Housing Investment in Four European Countries', PhD thesis, De Montfort University, Leicester.

Turner, P. (1989) 'Investment: Theory and Evidence', *Economic Review* 6, 3.

United Nations (1993) *Annual Bulletin of Housing and Building Statistics for Europe and North America*. Geneva: Economic Commission for Europe.

4

UNCERTAINTY IN EUROPEAN HOUSING MARKETS

Michael Ball and Michael Harloe

This chapter surveys international changes in housing provision over the past decade and is stimulated by four major concerns.

1   Policy makers believe that they are pursuing the right course by reducing housing subsidies and intervention. This is factually inaccurate – subsidies have been redistributed not reduced. It also ignores the strong arguments for state involvement in housing, even in market-led systems.
2   Insufficient regard is being paid to institutional change and uncertainty in the current era. Interest in institutions is only aroused when they fail to achieve policy objectives.
3   Compared to the 'golden age' era of the 1950s and 1960s, housing markets have become less stable but the implications of greater volatility are unclear. We want to identify here some of the serious consequences for households, housing costs and housing supply; ones which, we feel, justify state action to ameliorate market failure.
4   It is easy to justify cutbacks in state involvement in housing by blaming the need to reduce public expenditure in order to conform to Maastricht and monetary union criteria. It is similarly easy to explain housing changes in the same way. EU convergence factors are important (Ball and Grilli 1997) but here we will stress longer processes that have been generating structural change in European housing provision. To this end, considerable reference will be made to the USA, with its very different housing market traditions, in order to highlight in a comparative way some underlying themes that are helping to create the housing dilemmas faced by Europe today.

A decade of boom, crisis and weak recovery

The past decade has been tumultuous for housing in many countries. Owner-occupied markets boomed in the late 1980s then slumped – plunging financial institutions into crisis and causing large-scale mortgage defaults and repossessions

(OECD 1992). Countries which felt the worst effects of the property boom experienced long recessions in the early 1990s, and had slow recoveries in which housing trailed other sectors of the economy. Welfare safety nets have been strained to breaking point (economically and/or politically). Homelessness has changed from being a marginal to a major phenomenon. In 1991, for example, between 3 million and 5 million people were homeless within the European Union (Quilliot 1991). A moderate estimate for the US was that on any night 600,000 were homeless and 1.2 million over the course of a year (Burt 1992, cited in Dreier and Atlas 1995: 250–1). Household numbers are rapidly increasing in most countries as well, putting extra pressure on scarce housing resources. Despite these rising needs, most governments have reduced investment in low cost housing.

Political change, new market boundaries and old methodologies

The overwhelming governmental housing policy preference in the mid-1990s is for a private, weakly regulated market framework, which in the EU in particular marks a major ideological shift. For many years housing was regarded as a component of the welfare state, although with a large role for the market. In Western Europe (less so in the USA) it was treated as a social asset requiring special management and policy skills, extensive state regulation and subsidy. Much housing research involved the investigation of policy systems and subsidies to evaluate their effectiveness and efficiency. Relations between the state, the market and the housing consumer were shaped by politically determined factors.

Now, housing is primarily regarded as an economic asset. The belief in the efficacy of the market, with a subsidiary role for the state, has become part of the mainstream of political and policy thinking. This does not imply a return to the situation prior to major state intervention. This is not just because mass home ownership has supplanted mass private renting, with homeowners a more politically effective force than private landlords. More importantly, housing market outcomes are the result of continuously evolving processes which throw up new issues and problems. So governments cannot simply ignore housing by returning it to the unfettered market. Instead, there is now a powerful shift in the nature of the state–market interrelation. The scale of the shift and its unforeseen consequences mark out the 1990s as a climacteric for housing provision in the advanced economies.

The ideological shift back to the market has created two important dilemmas. The prime questions for policy centre on what constitutes effective housing markets, and how can their requirements be reconciled with social aspirations and political reality? The problem for housing studies is to discover the methodological tools, and the theoretical and empirical knowledge necessary to understand these markets. Especially in Western Europe, housing research is not geared up to understand current market mechanisms and state–market interactions because its main concern has been to examine a system of state intervention which is now being superseded.

Housing in many respects is like other goods or services. In housing markets buyers and sellers come together and equilibrium prices and quantities are set on the bases of the constrained preferences of consumers and profit-making drives of suppliers. Yet, housing also has some particular characteristics – it is very expensive, locationally specific and dependent on local environmental and other public goods. But these are only differences of degree, which is why many economists and New Right ideologues argue for minimalist housing policies.

Many markets, including housing, diverge from the simplistic competitive model. All forms of housing provision are associated with particular sets of social agencies – housing consumers, property owners, financiers, builders, land developers and owners and so on, and each of them has distinct interests and constraints. Their interactions fundamentally influence market outcomes. The specificity of market relations in housing supplies a continuing rationale for housing policies and for housing studies. Our argument, developed elsewhere (Ball *et al.* 1988; Ball and Harloe 1992), is that housing in both its market and non-market forms needs to be understood in the context of nationally and temporally specific, empirically grounded structures of provision rather than by ahistorical abstract ideal forms – like the stylised competitive market described above.

In the case of housing, the policy shift towards greater roles for market mechanisms has frequently occurred in ignorance of the specificity of its markets and the institutions active in them. The movement to a more market-led system has entailed changes in pre-existing housing agencies. Usually they have had to be changed in a piecemeal way, rather than simply being replaced with an ideal market blueprint. The fact that specific types of institutions, with particular interests and constraints, were pre-existent has forced governments into pragmatic and sometimes illogical deregulation and privatisation. This may explain why these changes have failed to deliver significant improvements in housing provision or even reductions in public expenditure.

Reorientation of housing provision has also had some unique features when compared to other recent market-based reforms. When European countries, especially the UK, privatised other public services and assets new market structures and regulatory bodies were invented, often after debate about the forms those markets should take. As economic agencies' behaviour depends on the incentives and constraints facing them, the new markets had to be planned to ensure that the new private sector suppliers achieved the desired objectives. In Britain for example, gas, electricity, water, public transport, telephones, telecommunications, television, and the 'internal market' within the National Health Service all have had markets invented by government for them. Conversely, there has been almost no concern about the forms that housing markets should take or the degree of regulation necessary for their effective functioning.

Policy and politics are not the sole reasons for changes in housing provision. Dynamics within systems of housing provision also create change, and generally policy is reactive to those changes. This claim is based on empirical observation rather than theoretical generalisation. Housing policy could involve revolutionary change, abolishing institutional structures and replacing them with new ones. This is currently occurring in some former Soviet bloc countries but in Western Europe and North America change is, as ever, evolutionary (Harloe 1995; Andrusz *et al.* 1996).

The most obvious structural change in many countries has been in mortgage finance. Most of the pressures causing the deregulation of the 1980s arose from general developments in financial markets and the specific problems of traditional mortgage lenders, so governments were under immense pressure to introduce reforms. Even so, the specific policy changes were influenced by prevailing policy debate. Serious miscalculations were made when drawing up the legislation: about the behaviour of financial institutions in the new deregulated environment; the willingness of consumers to take high borrowing risks, or their ignorance of those risks; the short-run adjustment impact on housing markets; and the effects of financial deregulation on national economies and the international economy (Kennedy and Andersen 1994). Hindsight makes all of us powerlessly wiser about the past. Yet few impact studies of the effects of deregulation were done (a partial US exception is Kane 1985). The lesson still does not seem to have been learnt – housing policy changes remain driven by beliefs rather than analysis.

Economic integration

Financial deregulation illustrates another characteristic of the current dynamics of structures of housing provision. The emphasis on market-based reforms has been associated with an increased penetration of housing structures of provision by institutions in the wider economy. One effect is to make housing provision more dependent on general macroeconomic performance. Again, the impetus for change has come as much from institutions as from governments but, nevertheless, the impact on policy is striking.

A central tenet of earlier welfare-state oriented housing reform was that people's housing conditions should not be completely determined by their or the nation's economic circumstances. Obviously this de-coupling had limits but it was extensive, being exercised through subsidies, tax reliefs, regulated interest rates and rents, heavily subsidised programmes of slum clearance, house-building and renovation, low rent social housing, etc. These interventions created a special, protected space for housing institutions and consumers. This is now in dissolution. Specialist housing institutions are disappearing or being reshaped on a market model; regulated housing costs are vanishing; mortgage interest rates are market determined; subsidies are switched from 'bricks and mortar' to 'people in need'. Housing is increasingly dependent on the functioning of the wider economy and relative levels of housing consumption are determined by people's positions within it.

Some unexpected consequences have resulted. For example, general consumers' expenditure is increasingly influenced by net housing wealth (Miles 1994). Housing markets have increasingly fluctuated in phase with the general economy, reinforcing rather than counteracting macroeconomic cycles. The housing tail ends up wagging the dog called the economy. Effects are transmitted internationally, increasing the scale of fluctuations in the world economy. Housing investment cycles, for example, used to help balance out fluctuations in individual countries' macroeconomic cycles; nowadays they are a transmission mechanism for international demand fluctuations (Ball and Wood 1994).

Housing is also strengthening regional economic disparities. The negative equity of homeowners, for instance, affects aggregate demand in a region, so that previously buoyant areas, like south-east England or the Boston area, experienced exceptionally depressed consumer demand and housing markets in the early 1990s (Nationwide Building Society 1995; Forrest and Murie 1994; Case 1992). Other changes go beyond amplified business cycles. Economically depressed regions now have depressed housing markets to contend with and far less regionally improving, housing-related, subsidised urban renewal. Conversely, the move towards income-related rental subsidies puts more money into thriving regions with their higher rent markets. These subsidies may also exacerbate regional rent differentials through tenant/landlord collusion in rent setting (Hills 1993; Kemp 1992). In addition, in overheated regional labour markets, housing shortages discourage inward mobility because rising demand is no longer catered for by government actions.

In some countries large-scale migration, caused by exogenous factors, has had to be accommodated but unaided markets have not been able to respond adequately. An example is the movement from Eastern Europe into western Germany, which forced a temporary expansion of social rented housing production, reversing former policies (Harloe 1995). There is also the mass migration from Asia and south-central America into the USA. Of course, the housing problems of low income, non-white migrants have long been issues in the US – with problems often attributed to insoluble 'market failure'.

Impact on consumers

The rising cost of housing

The most obvious consequence for consumers is that housing is becoming more expensive over the medium term, even when cyclical effects are discounted. In Western Europe the sharpest rise between 1975 and 1987 occurred in the Netherlands, where housing costs as a percentage of total household expenditure rose by 30 per cent. In France, West Germany and Denmark the rise was around 17–20 per cent and in the UK 11 per cent. By the late 1980s the figure was 18–20 per cent in most Western European countries but was 26 per cent in Denmark (due to very high interest rates) (Boelhouwer and van der Heijden 1992; Myers et al. 1992; Bramley 1994). These data illustrate the point that, as in the USA, housing circumstances are now far more closely correlated with a person's economic circumstances than previously.

Of course, most households are better housed now because average living standards have risen. But as income distribution has become less equal in most countries, so have individuals' housing standards. As poverty has risen, so have poorer housing conditions and problems of access and affordability.

Inequality in housing outcomes

The association between changes in incomes and housing conditions is imperfect. Some groups gain more from housing subsidies than others, so that their housing conditions are better than they would otherwise be. There is also a disjunction between the life cycle of earnings and individuals' housing costs and standards. For example, in Britain, people often become homeowners in their twenties, so their housing costs are high two decades or more before their earnings peak. A partial solution to this life cycle problem is for households to trade up when they can afford to. This results in a temporal profile of housing costs relative to incomes that consists of a series of peaks at times of moving, with gradual declines until the next move up or to when the household eventually trades down, releasing housing equity. In such a system, with large amounts of trading up and down, the ability of households to get the timing of moves right requires forecasting skill and there is considerable risk and uncertainty. Home-owner moves bunch during market upturns, adding to the market volatility, which raises uncertainty. With greater risk, there will be a greater array of winners and losers in the housing market. The systemic risks generated were obscured in the 1970s and early 1980s because post-subsidy real mortgage interest rates were often negative, so the financial risks faced by homeowners were low.

Other tenures can generate similar disparities. In many Western European countries there is a disjunction between earnings and social rented housing costs. This occurs in systems like the German or Danish ones, where rents are set on a historic cost basis (plus management and maintenance costs). Established tenants with higher incomes often occupy units with low rents. Meanwhile, new, lower income entrants occupy costly new units.

Access

Another generator of disparities between incomes and living conditions is the increasing difficulty of getting access to housing. In Western Europe production cutbacks and, in some cases, privatisation mean that social housing is increasingly available only to certain low income groups. Access to home ownership is impeded by substantial entry costs and high real levels of initial mortgage repayment. In some cases negative equity has locked homeowners into devalued properties.

A market-based response has been to lower housing standards for entry level dwellings (Karn and Sheridan 1994). In Britain, legal minimum social housing standards were abandoned in 1979, and later also in the Netherlands. In Germany, however, the high standards of new housing, and hence high barriers to entry, were sustained. Whether a decline in standards is a rational response to a temporarily acute problem or a chronically short-sighted strategy that imposes higher long-term costs is debatable.

Vulnerability

The outstanding characteristic of contemporary housing provision is how vulnerable housing circumstances now are to sudden changes in the variables against which households calculated their expected housing costs and their ability to pay. This is clearest in owner-occupied housing, where sharp rises in interest rates were a prime cause of mortgage default in the late 1980s and early 1990s. Moreover, unemployment or the sudden breakdown of a relationship can plunge households into debt and housing distress. Such vulnerability has always been greater in the US housing system. But, previously, high levels of employment for the white majority, rising real incomes and long-term fixed, below market-clearing, mortgage interest rates were sources of considerable housing security. They are not so now.

Generally, broader economic and social changes are rapidly increasing the number of households vulnerable to housing distress. In Europe, after each of the major recent recessions (mid-1970s, early 1980s and late 1980s/early 1990s), unemployment did not fall to pre-recession levels. Job security has been permanently reduced for growing sectors of the population. Economic and labour market changes are resulting in an increasing polarisation of incomes and living conditions and more uncertainty, not just for the poor but for the 'middle mass' too. In this new situation, with regard to housing, there are only limited policy strategies to deal with unexpected calamities hitting previously stable and economically secure households. Certainly the obvious market solution, private insurance, cannot do the job effectively – for adverse selection and moral hazard reasons, if for no other.

Consequences for housing supply

House-building is an industry in decline. A common Western European pattern was for output to peak in the late 1960s or early 1970s (when it was also high in the USA). Since then there has been a secular decline. In Europe only the Netherlands, with very rapid growth in households, sustained production in the 1980s. In 1989 West German output was 45 per cent below the 1975 level, in France the fall was 37 per cent, in Denmark 34 per cent and in the UK 27 per cent. Recovery from the early 1980s recession was weak and output fell sharply as the early 1990s recession set in (Ministry of Housing, Physical Planning and Environment 1991; Boelhouwer and van der Heijden 1992), although by the mid-1990s recovery was stronger. Social housing completions have varied considerably. In some cases there was continuous decline. But in the Netherlands, Germany and Denmark governments sometimes increase output for macro-economic or political reasons. These were temporary revivals, not a reversion to large-scale social housing to meet rising housing demand.

Britain illustrates house-building decline the most starkly. Housing output has fallen for over two decades with output now only about a third of that during the peak years (Ball 1996a). This means that the house-building industry has contracted as sharply as many of the older manufacturing industries in Britain, a fact ignored in debates about the de-industrialisation. None of the standard reasons for de-industrialisation – large-scale productivity increases, shifts in patterns of demand, or intensified foreign competition can explain the loss of housing output and jobs. The basic problem is that, unlike most other major consumer durables, the real cost of building new houses is rising sharply. So private house-building has problems sustaining its own output rather than taking over as public output declined, as governments had hoped.

Some argue that the output decline reflects demographic change and a much needed reduction in subsidies, and that the existing stock can be refashioned to provide high space and quality standards. Yet such arguments ignore the obvious pressures on housing demand in the medium term. Meanwhile the rapid growth of single-person households and other changes resulted, between 1970 and 1987, in almost a 50 per cent growth in households in the Netherlands, 30 per cent in France and around 22 per cent in West Germany, the UK and Denmark (Ball and Wood 1994; Boelhouwer and van der Heijden 1992).

Recent work on housing demand in England for the period 1991–2011 suggests a need for an average annual output of 240,000 (including 90–100,000 units of social housing). In the early 1990s actual output was around 150,000, with social housing output barely a third of the required level (CSO 1995; Joseph Rowntree Foundation 1995).

Governments and consumers are ever more reluctant to pay for increases in the housing stock, despite the supposed fact than the long-term demand for housing rises roughly at the same rate as the growth in personal incomes. On this basis, we should expect a real rise in housing demand of 15–25 per cent a decade at typical GDP growth rates; though relative prices, tenure shifts, demographics and labour markets complicate the picture. How will the supply side cope in future – just with price rises or (hopefully) through extra output?

Two concerns for housing policy arise from the supply side:

1   Much effort has been put into identifying the distorting effects of subsidies on housing demand and tenures, but there is less recognition that the supply response of house-builders to subsidy-inflated house prices has been poor and is getting worse. The standard response is to blame inadequate land supply but the limited evidence indicates that this is not the only cause. Problems with other building inputs, the efficiency with which they are used

and low technical innovation are important contributory factors (Ball 1996b).

2 There are well-publicised debates over whether governments can afford the public expenditure to maintain good health care, pensions and other social provision. A similar question can be asked of housing. As demand for more and better housing grows, while real costs rise, prices or subsidies have to continue to grow to induce more output. Who will pay and how? Will it be done through higher housing consumption costs, higher taxes to pay for extra subsidies, or with poorer than desired housing conditions for growing numbers of people?

There is insufficient research on the structure of the house-building industry, how it is responding to decline, and how it would cope with a major revival of demand. Industrial decline is accompanied by greater fluctuations in demand which are probably retarding long-term productivity and efficiency. We do not know what are the effects of market decline and greater market uncertainty on the house-building industry. Other great unknowns include the functioning of the land market. In some countries the activity of the financial sector in land speculation is well documented, for example Japan (Oizumi 1994). But quite what encourages landowners to enter the development pipeline, the nature of the intermediaries involved and cross-national institutional differences are all under-researched.

House-builders and developers are major users of planning systems and they encourage landowners to apply for planning permission. Debates over planning should consider the roles of these key agencies. Yet there is little information on how the residential land market works, how landowners, builders and planners interact, whether 'best-practice' procedures can be identified and copied or if the systems need fundamental reform.

We raise these points for three reasons.

1 To reiterate that if housing analysis wishes to move beyond simple market models it must take account of many poorly understood aspects of structures of housing provision.

2 New structures of provision raise old problems in new ways. Conflicts over greenfield land for new housing, for example, have been major sources of contention for decades. The solution in Europe has generally been to have restrictive planning with large-scale public sector infrastructure expenditure. In the USA the solution has depended on a high consumption of energy in car transportation and large-scale land transfers to suburbanisation. In Europe, planners will be hard pushed to sustain land constraint in the face of more market-driven housing suppliers, whilst they face opposite pressures over the environment.

3 Housing policy innovations often miss important supply side links. For example, the construction industry might play a role in a revival of the

privately rented sector, as it did in the nineteenth and early twentieth century (for Britain see Dyos 1961).

Housing on both the demand and supply sides is becoming a more uncertain business. A key role for housing policy could be to reduce uncertainty in order to increase investment and lower costs. Yet recent policies have heightened uncertainty.

The end of the owner-occupied dream

A central tenet of housing policy in Western Europe is that most households want to be, and should be able to be, owner-occupiers. The recent housing booms and busts have severely dented this belief. In an era of low inflation, high real interest rates, reduced tax reliefs and volatile housing markets, owner-occupation is no longer a clear one-way bet.

In the 1970s, when the real cost of home ownership was very low and real incomes were rising, its share of the total housing stock grew rapidly – by 7 per cent in the Netherlands, 6 per cent in the UK, 5 per cent in Denmark and 2 per cent in France and Germany. In the 1980s the picture was far more mixed – the share declined by 1 per cent in Denmark, was static in Germany and grew far more slowly, by 3 per cent, in the Netherlands. However, France made up for its earlier slow growth with a rise of 7 per cent and in Britain there was a remarkable growth of 10 per cent, so that by 1990 the homeowner share, at 65 per cent, was above the American level. However, in both these latter cases, growth was powered by large-scale state intervention and expenditure. In France these involved expensive (and unsustainable) subsidies for new homeowners. In Britain there were massive asset transfers (in fact, the largest privatisation in this period) through the sale of council housing to tenants at discounted prices. A similar pattern occurred in the USA where the home ownership rate rose from 63 per cent in 1971 to 67 per cent in 1980, but then fell back throughout the period up to 1993 (Ministry of Housing, Physical Planning and Environment 1991; Joint Center for Housing Studies 1991; Karn and Wolman 1992; US Bureau of the Census 1994).

Home ownership is inevitably an uncertain venture. It involves long term repayment profiles which have to tie in with life-cycle earnings, interest rates have to be forecast, and employment stability is vital. In addition, the future capital value of the property has to be forecast. Real house prices fluctuate substantially at national levels and even more so locally, so not only the long-term but also the cyclical price trends must be taken into account. Finally, households have to consider alternative housing and investment options. This complex calculation is difficult and many, not surprisingly, get it wrong. The market-driven reforms in housing during the 1980s made the calculation more difficult and the outcomes more risky.

The boom years of the 1980s occurred when many governments aimed to increase owner-occupation in Western Europe by making social housing more expensive and less available, and through financial liberalisation. However, liberalisation, by weakening the traditional rationing mechanisms of mortgage finance without replacing them with consumer-oriented risk assessments, helped to draw into home ownership households for whom the risk is too high, whilst reducing alternative housing opportunities if those risks turned sour.

Boom and bust cycles are endemic to owner-occupied housing markets, particularly if supply elasticities are low, which they are in most EU countries. But, every time there is a bust, soothing words emanate from financial and owner-occupier lobbyists, 'Don't worry, the lesson has been learnt, it will never happen again', which in Europe in particular has distracted policy makers from thinking of ways to reduce housing market volatility.

Although rationed mortgage finance systems contain inefficiencies, they do effectively limit over-optimistic consumer ambitions. Mortgage rationing takes two forms: creditworthiness criteria and quantity constraints. In the past, the two interacted. When there was chronic excess demand, lenders became more selective. Under rationing, excess demand was encouraged because the stringent lending criteria led to lower mortgage defaults and therefore to reduced-risk premiums in mortgage interest rates. The low risk of the business encouraged savers to accept lower interest rates. This, plus their quasi-monopoly market positions, enabled Western European lenders to price their mortgages at below market clearing rates. The position differed in the USA but a similar situation was created through federal regulation (Ball 1990).

Deregulation in mortgage finance is often cited as a major cause of house price booms in the 1980s because financial institutions misjudged the security of mortgage lending or blatantly ignored prudent lending criteria. In the early 1990s, lenders were more wary, so creditworthiness rationing was reinforced. Memories then faded and personnel changed with the progression of time, and competitive pressures grew in some mortgage markets, so relaxation of lending criteria began all over again – and by 1997 house price booms were well under way in some markets, such as in London.

Housing markets have not reverted to being a low risk lending opportunity, so risk premiums on mortgage interest rates are higher than they were a decade ago. Furthermore, mortgage institutions have lost many of the privileges that they had which helped to keep mortgage interest rates low. Mutuals have voluntarily 'privatised' themselves in the UK and USA (Ball *et al.* 1997). Consequently, discounting short-term fluctuations, the relative price of mortgage finance has risen significantly. Thus, other things being equal, *fewer* rather than more households can invest in home ownership.

Mortgage liberalisation is a prime example of a widely adopted housing reform which produced effects in direct contradiction to stated aims of reducing borrowing costs and increasing home ownership. Other recent changes, such as reduced tax reliefs and higher property taxes, have had similar consequences for the tenure. It is reasonable to conclude, therefore, that governments frequently have failed to understand the market logic of the housing reforms they have implemented.

An important cause of current problems in owner-occupied housing markets is the working through of market adjustments to the higher risks generally involved in the tenure. Marginal owner-occupiers, earlier encouraged into the tenure, are being forced out; while some households that previously would have entered the tenure are now better off renting. Some households are now better off buying a home later in their lives than would have been the case in the past. Meanwhile, rental housing supply is not responding to the altered pattern of owner-occupier demand.

The failure to redress the decline of rented housing

Continuing decline

The decline in private rental housing in Western Europe and the USA has been evident over many years (Harloe 1985). The current conditions of risk and uncertainty have not encouraged any upsurge in private market investment. If this is true for home ownership, it is even more so for the private rental sector, which accommodates many sections of the population whose economic prospects have become increasingly uncertain. Therefore, it is not surprising that new investment in private rented housing has continued to fall in most Western European countries (van der Heijden and Boelhouwer 1996).

Owner-occupied dwellings converted to rental

Historically, tenure transfers mainly have been one way, with rental units converted to home ownership. There are indications that the 1990s may be seeing a reverse movement and therefore a slowing down of decline or even a slight reversal in private renting's share of the housing stock. There are several reasons for this, almost all related to conditions in the homeowner market. Some housebuilders have rented properties built for sale, due to lack of demand from buyers; some mortgage lenders have rented repossessed properties, rather than have them stand empty; and hard pressed mortgagors have sub-let rooms or rented their dwellings to meet loan repayments, or have been forced to rent rather than sell their properties when job changes or other exigencies make a move essential.

Private rather than social renting

Some governments have tried to encourage a revival of the private rented sector as a substitute for social rented provision (van der Heijden and Boelhouwer 1996). Revival requires a major extension of investment subsidies, cutting across the general desire to reduce them. So, again, there is a contradiction between a housing policy objective and wider economic and fiscal policies. In Britain, the

Conservative governments wished to revive private rental housing as an alternative to the social sector. When rent deregulation failed, they were reluctant to provide large-scale subsidies. Yet analysts agree that only large-scale and continued state subsidies would enable a real revival of private rental investment to occur (Crook 1993; Crook and Kemp 1996).

Germany is a special case. Here the demands arising from reunification and migration from the East have led to rising rents and temporary government action (by increasing depreciation allowances) to stimulate private sector production, so by 1992 46 per cent of new output was in this sector (van der Heijden and Boelhouwer 1996).

Social shyness

As we have indicated, governments have also been unwilling to make major new commitments to subsidising social rented housing. This is why, bar some temporary exceptions, social housing output has been in decline. The key objective has been to reduce investment, to raise rents nearer to market levels and to encourage greater reliance on the private sectors and on the household's own resources. A central feature has been the progressive restriction of state support to income-related housing allowances. The idea was that most tenants would pay higher rents and smaller sums could be paid in subsidies for the poorest households only. (This would also encourage a shift to home ownership.) However, this strategy has gone badly wrong, as unemployment and low incomes among social tenants have escalated, along with the cost of housing allowances (Papa 1992).

A recent study of the rising costs of housing support in Britain, France and Germany shows that the sharpest rises in housing allowance costs have occurred in Britain, which has moved away from social insurance towards targeted, means-tested benefits. By contrast, in France and (even more) in Germany the rising cost of support for housing expenditure due to unemployment and poverty has been absorbed by more generous unemployment and other social insurance benefits. In all cases, though, policies which have encouraged higher rents have led to higher welfare state expenditure to meet such costs (Evans 1996).

Several governments have attempted to privatise the social housing stock. The best-known case is Britain, with the large-scale sale of council housing to its tenants from 1980, followed later by partially successful attempts to transfer whole projects to the private sector, housing associations or tenant ownership. Council housing's share of the total stock fell by about 25 per cent in the 1980s as a result of these policies.

In Germany, large-scale privatisation has also occurred but differently. Much of the German social housing stock is owned by private investors. The commitment to keep it in the social sector only lasts as long as the owners are repaying their state subsidised loans. Changes in the early 1980s (by a Social Democrat–Free Democrat government) allowed repayments to be accelerated and, hence, enabled owners to make capital gains by selling out, or to raise their profits by charging market rents, as the properties joined the private rental sector. A second major change occurred in the late 1980s, when the tax privileges that the limited-profit social landlords enjoyed were abruptly removed (except for some mainly middle income co-operatives). Formerly, these entailed that this social housing was permanently subject to regulated rents and access. Now these (ex)social land-lords are also free to join the private market when their state loans are repaid. In 1984 there were around 4 million social rented dwellings in former West Germany, but, according to one estimate, by 2000 there will only be around 1 million (Haussermann 1991).

Neither the Dutch nor the Danish governments have been as able or willing to privatise social housing. In both countries this housing continues to accom-modate many middle income households, although polarisation is occurring. It retains, accordingly, considerable popular and cross-party support (Harloe 1995). As in the United States, legal factors also inhibit privatisation. However, both governments share the desire to reduce state engagement in the sector. So they have pursued a more subtle form of privatisation – a return to an earlier conception of non-profit social housing as (to use the Dutch phrase) a 'private initiative', outside the public sector, with minimal reliance (except for housing allowances) on state subsidies and greater dependence on internal financing of new investment. A particularly vigorous start has been made along this new path in the Netherlands, where major housing policy reforms were introduced in the late 1980s and early 1990s (CECODHAS 1994).

Only France is an apparent exception to this general policy shift. One reason is that alone of all the countries mentioned here it had a socialist government throughout most of the 1980s and in the early 1990s. The Socialist Party had come to power pledged greatly to increase the output of social rented housing. But by 1982 it had been forced to abandon this and subsequently maintained a modest level of social rented housing output by drastically reducing subsidies for social home ownership. But there was no move by any of the major political parties to advocate privatising the existing social rented housing stock. More recent centre-right administrations have been cautious in changing previous housing policies. Indeed, there was new support for the housing sector in 1993 in order to raise sagging output to above 300,000 (Kleinman 1995).

Conclusion: a new focus for housing policy?

In an earlier study (Ball *et al.* 1988) we outlined five key changes in housing markets and policies: the decline in house-building and the rise in the real cost of house purchase; the negative effects of mortgage market deregulation; a far more difficult environment for the producers of housing; the trend towards the residu-alisation of social rented housing, both in terms of new supply and in relation to its social and economic circumstances; and the general desire of governments to disengage from housing provision.

This analysis still seems broadly valid a decade later, even though many changes have occurred. Governments have continued to make radical shifts in the boundaries between the state, the market and the individual household and hence in structures of housing provision. The manner in which they have done this has had some negative and even self-defeating consequences – partly because of the inherent incoherence of some policies and partly because the structures of housing provision which they are trying to influence have been subject to other pressures and have altered in content and composition. As we have argued, risk and uncertainty have been increased for the very private sector investors that governments now wish to rely on. Equally, risk and uncertainty have been increased for housing consumers. The cost of housing has continued to escalate and supply and accessibility have been constrained.

Nothing suggests that private housing markets are working as many economists (and politicians) believe that they should. Attempts to free up the private market and reduce state interference have failed to produce the desired results. Meanwhile, many housing consumers have been unable to find any housing at all, or have lost what they did have; many more have been trying to cope with unaffordable housing payments; others have been unable to progress up the housing quality ladder, or to relocate for better job opportunities.

If greater market emphasis is here to stay, how might policies relate to it in more advantageous ways? The most obvious answer is the avoidance of ill thought out attempts to 'get back to the market' by deregulation. There is also growing recognition that policies which simply force up individual housing costs to insupportable levels have negative consequences, that they shift the subsidy burden between forms of assistance rather than reducing it substantially, and they cause homelessness, economic hardship and reductions in labour market efficiency.

There is an urgent need for housing policy (and housing research) to be based on an effective analysis of its wider social and economic, as well as its housing market consequences. New justifications and policies for state intervention in housing are needed. Much debate has centred on how governments have got their interventions wrong rather than how they could get them right. In part, the negative attitude to state intervention is a product of ideological fashion but it is also a consequence of focusing on tenures rather than on housing provision as a whole.

An analogy can be made with changes in the intellectual discussions concerning the economics of the environment. The economic costs of banning or minutely directing all pollution emissions (total statism) or, alternatively, of *laissez-faire* market solutions are both now seen as too radical to contemplate, and they both raise uncomfortable democratic and distributional issues. Instead, reasons and principles for effective environmental state intervention have been elaborated, involving the relevant economic theories. No consensus has emerged but the issues are now well understood. Once a broad intellectual framework was accepted dialogue was possible. In like manner, housing debate needs to move away from a dispute between the state versus the market. Justifications for state intervention have to be rethought and the potential conflicts between housing market efficiency, distributional equity and wider social and economic problems have to be identified and continually re-evaluated. Only then do tenures enter the discussion as part of structures of housing provision, to which the benchmark theoretical criteria can be applied.

In the muddy real world, conflict between housing market goals and other policies also has to be recognised. An internally efficient and consistent housing system may be at variance with, say, the functioning of a local labour market or national economic development. So toleration of housing market inefficiencies, for example, may be preferable to chronic labour supply shortages or to periodic speculative property market booms.

There are good reasons for governments in democratic societies to be concerned about housing provision. On the demand side, despite many similarities with other commodities, housing is still distinctive. It is important to people not only because of its expense but also because of its central part in their lives – economic, social and psychological. Governments also have an intergenerational obligation. Children have little independent choice over their housing circumstances but housing conditions in early life, in association with other social factors, are a major determinant of later abilities and opportunities. On the supply side, housing markets exhibit characteristics of 'market failure'.

Politicians have long recognised the special place of housing in people's lives, and, with varying degrees of understanding, housing market failure. Unfortunately the knee-jerk response has often been to espouse particular forms of provision. It would be better to get our principles right first, and then see structures of housing provision as delivery systems which, within limits, can be steered in particular ways. After all, housing tenures and markets are means rather than ends. Europe has a long tradition of innovative housing policy – a tradition that now needs to be put to good use.

References

Andrusz, G., Harloe, M. and Szelenyi, I. (eds) (1996) *Cities after Socialism*. Oxford and Cambridge, MA: Blackwell.

Ball, M. (1990) *Under One Roof: Retail Banking and the International Mortgage Finance Revolution*. Hemel Hempstead: Harvester Wheatsheaf.

Ball, M. (1996a) *Housing Investment: Lessons for the Future*. Bristol: Policy Press.

Ball, M. (1996b) *Housing and Construction: A Troubled Relationship*. Bristol: Policy Press.

Ball, M. and Grilli, M. (1997) *Housing Markets and European Convergence in the EU*. London: Royal Institute of Chartered Surveyors.

Ball, M. and Harloe, M. (1992) 'Rhetorical Barriers to Understanding Housing Provision: What the "Provision Thesis" Is and Is Not', *Housing Studies* 7, 1: 3–15.

Ball, M. and Wood, A. (1994) 'Housing Investment: Long-run International Trends and Volatility,' Birkbeck College Discussion Papers in Economics, 11/94, London.

Ball, M., Harloe, M. and Martens, M. (1988) *Housing and Social Change in Europe and the USA*. London: Routledge.

Ball, M., Kenway, P. and Palmer, G. (1997) *Housing Risks and Opportunities: Reforming Mortgage Finance*. London: New Policy Institute.

Boelhouwer, P. and van der Heijden, H. (1992) *Housing Systems in Europe, Part I. A Comparative Study of Housing Policy*. Delft: Delft University Press.

Bramley, G. (1994) 'An Affordability Crisis in British Housing: Dimensions, Causes and Policy Impact', *Housing Studies* 9, 1: 103–124.

Burt, M. (1992) *Over the Edge: The Growth of Homelessness in the 1980s*. New York: Russell Sage Foundation.

Case, K.E. (1992) 'The Real Estate Cycle and the Economy: Consequences of the Massachusetts Boom of 1984–7', *Urban Studies* 29: 171–183.

CECODHAS (1994) 'Netherlands. "Brutering": The Plan to Make Social Housing Organisations Independent', *Observatoire Européen du Logement Social* 6 (June). Paris: UNFOHLM.

CSO (Central Statistical Office) (1995) *Annual Abstract of Statistics*. London: HMSO.

Crook, A. (1993) 'The Revival of Private Renting in Britain: Deregulation, Tax Incentives and Private Finance', *Housing Finance in the 1990s*. Gävle: National Swedish Institute for Building Research, pp. 273–288.

Crook, A. and Kemp, P. (1996) 'The Revival of Private Rented Housing in Britain', *Housing Studies* 11, 1: 51–68.

Dreier, P. and Atlas, J. (1995) 'US Housing Problems, Politics and Policies in the 1990s', *Housing Studies* 10, 2: 245–269.

Dyos, H.J. (1961) *Victorian Suburb: A Study of the Growth of Camberwell*. Leicester: Leicester University Press.

Evans, M. (1996) *Housing Benefit Problems and Dilemmas: What Can We Learn from France and Germany?* STICERD Welfare State Programme, WSP/119, LSE, London.

Forrest, R. and Murie, A. (1994) 'Home Ownership in a Recession', *Housing Studies* 9, 1: 55–74.

Harloe, M. (1985) *Private Rented Housing in the United States and Europe*. Beckenham: Croom Helm.

Harloe, M. (1995) *The People's Home? Social Rented Housing in Europe and America*. London and Cambridge, MA: Blackwell.

Haussermann, H. (1991) 'Housing and Social Policy in Germany' (mimeo), Bremen University.

Hills, J. (1993) *The Future of Welfare: A Guide to the Debate*. York: Joseph Rowntree Foundation.

Joint Center for Housing Studies (1991) *The State of the Nation's Housing 1991*. Cambridge, MA: Harvard University.

Joseph Rowntree Foundation (1995) 'Housing Demand and Need in England 1991–2011', *Findings*, Housing Research 157 (October). York: Joseph Rowntree Foundation.

Kane, E. (1985) *The Gathering Crisis of Federal Deposit Insurance*. Cambridge, MA: Ballinger.

Karn, V. and Sheridan, L. (1994) *New Homes in the 1990s. A Study of Design, Space and Amenities in Housing Association and Private Sector Housing*. York: Joseph Rowntree Foundation.

Karn, V. and Wolman, H. (1992) *Comparing Housing Systems. Housing Performance and Housing Policy in the United States and Britain.* Oxford: Clarendon Press.

Kemp, P. (1992) *Housing Benefit: An Appraisal.* London: HMSO.

Kennedy, N. and Andersen, P. (1994) *Household Saving and Real House Prices: An International Perspective.* Basle: Bank for International Settlements.

Kleinman, M. (1995) 'Meeting Housing Needs through the Market: An Assessment of Housing Policies and the Supply/Demand Balance in France and Great Britain', *Housing Studies* 10, 1: 17–38.

Miles, D. (1994) *Housing, Financial Markets and the Wider Economy.* Chichester: Wiley.

Ministry of Housing Physical Planning and Environment (1991) *Statistics of Housing in the European Community.* The Hague: Ministry of Housing.

Myers, D., Peiser, R., Schwann, G. and Pitkin, J. (1992) 'Retreat from Homeownership: A Comparison of the Generations and the States', *Housing Policy Debate* 3, 4: 945–975.

Nationwide Building Society (1995) *Housing Finance Review* 2 (July). London: Nationwide Building Society.

OECD (1992) *OECD Economic Outlook 52.* Paris: OECD.

Oizumi, E. (1994) 'The Japanese Property Boom', *Environment and Planning A* 26: 199–213.

Papa, O. (1992) *Housing Systems in Europe, Part II. A Comparative Study of Housing Finance.* Delft: Delft University Press.

Quilliot, R. (1991) 'Preface', *Marchés et politiques du logement dans la CEE.* Paris: La Documentation Française, 7–12.

US Bureau of the Census (1994) *Statistical Abstract of the United States: 1991.* Washington, DC: US Government Printing Office.

van der Heijden, H. and Boelhouwer, P. (1996) 'The Private Rental Sector in Western Europe: Developments since the Second World War and Prospects for the Future', *Housing Studies* 11, 1: 13–34.

5

THE IMPACT OF EUROPEAN INTEGRATION ON THE CONSTRUCTION INDUSTRY

Hugo Priemus

This chapter offers a critical analysis of the impact of European integration on the construction industries of the countries of the European Union. We draw a distinction between the consequences of European regulations and those of an integrated market within the European Union.

The direct and indirect consequences of European integration on earthworks, waterworks and roadbuilding have been examined by the *Vereniging van Boorondernemers en Buizenleggers* (the Association of Drilling and Pipeline Construction Companies), the *Nederlandse Vereniging van Wegenbouwers* (the Dutch Roadbuilders' Association) and the *Vereniging Aannemers Grond-, Water- en Wegenbouw* (Association of Earthworks, Waterworks and Road Building Contractors) (1991). Their conclusions on *direct* impacts are relevant to the whole of the building industry, and are reproduced below in Table 5.1.

The most important *indirect* consequences of the European single market for contractors in the construction industry are as follows:

1   It has become possible to obtain orders in other EU countries. The opportunities for contractors to win contracts abroad could expand further, both because of legislation with regard to tendering practices, and because of the gradual adoption of a more international outlook by clients in the countries surrounding our own. This will be a gradual process as local culture and nationalism are often deep-rooted considerations.
2   Domestic clients will start to look abroad for possible contractors. This will also be a gradual process, but it is likely to take place, especially in border areas and where large projects are involved.
3   The creation of a single European market is expected to generate market growth. This will also apply to construction-related markets, many of which, such as telecommunications and energy, remain quite closed and restrictive. However, the exact scale of this growth is hard to predict.

Table 5.1 Direct consequences of European integration for the building industry

Development	Consequences
Free movement of employees	It will be easier to attract foreign workers to places where the local workforce is insufficient
Harmonisation of diplomas and professional qualifications	It will be easier for workers to look for work abroad
Harmonisation and normalisation of technical norms and standards	National regulations will increasingly be attuned to European regulations
Liberalisation of telecommunications	Positive market influence A more complicated market
Liberalisation of energy	Market expansion in the long term
Relaxation of the publication laws. Fiscal harmonisation	Improved conditions for transborder operation
Liberalisation of financial services	It will be easier to choose the cheapest banking and insurance services
Shaped EU policy on bidding	Various existing agreements between contractors and between contractors and government will be subject to closer scrutiny
Social Europe	There will be more European legislation on employees' working conditions, safety and health
Liberalisation of governmental contracts	There will be more market possibilities abroad There will be more foreign competition for the larger projects
Responsibilities	The contractor is expected to be accorded wider responsibilities

Source: VBB, NVW and VAGWW (1991: 10–11).

Harmonisation of technical regulations and standards

Directives

In order to harmonise technical regulations as far as possible the EU has laid down a number of directives, which must be implemented by the member states by a given date. The arguments in favour of uniform technical regulations at the European level are many and may be summarised as follows (Priemus 1991):

- the promotion of interchangeability, compatibility and complementarity of components and products;
- the utilisation of scale effects;
- the exchangeability of information;
- the reduction of other costs (because, for instance, there are fewer conflicts about contractually agreed quality);
- greater freedom of choice for customers;
- greater freedom of choice for producers and contractors, by identifying and giving market access to equivalent products.

The approach that the European Commission has followed since 1985 has been twofold: to abolish technical barriers to trade within the EU (Pelkmans 1985) and to stimulate the formation of a European policy in the fields of health, safety, consumer protection, employment and the environment. The extensive technical appendices that characterised the older, less successful approach have been abandoned in favour of harmonised technical regulations that establish minimum safety and health standards. It is clearly easier for member states to agree on minimum standards than to agree on every detail.

The implementation of the single market will require around 300 EU directives to be incorporated by member states into their national legislation. Meanwhile a number of directives and draft directives have already been drawn up that are of immediate importance to the building sector (Langeveld and De Vries 1990):

1 The *Co-ordination Directive on Government Tendering*. After a first version in 1971, a draft directive was submitted in 1986 to the Council by the Commission. This directive relates to the co-ordination of procedures for inviting tenders by governments or government bodies. This directive was amended on 18 July 1989.

2 The *Building Products Directive* was adopted by the Council on 28 June 1988 (Walters 1988: 66). It relates to the amendment of the legislative provisions of member states concerning products intended for the building trade, and outlines requirements relating to health and safety. It is of great importance to the building and building supply industries.

3 The Directive concerning the *mutual recognition of degrees, diplomas and other qualifications in the field of architecture* includes measures to facilitate the exercise of the right to set up a practice. The directive, adopted in 1985, has meanwhile been incorporated into the Architects' Title Act in the Netherlands, and so is of great importance to architects and architecture courses. A directive similar to that for architects is in preparation for *engineers*.

4 The *Supplies Directive* was adopted by the Council in December 1976 and revised and amended in 1980. The directive relates to the co-ordination of procedures for placing government contracts. A draft proposal to amend the directive further was submitted to the Parliament by the Council in December 1986.

5 The *Information Directive* relating to technical regulations and standards had already been adopted in mid-1983. The clauses concerning technical regulations came into effect in 1984, and those concerning standards in 1985.

6 The *Product Liability Directive* dates from 1985. We shall return to this later.

The placing of government contracts is subject to the *Liberalisation Directives* (1971) and the *Co-ordination Directives* (1971/1989), which determine the lifting of limitations and the co-ordination of procedures, respectively. The

'Services', 'Work' and 'Deliveries' directives are also important, since they all have the same purpose: identical conditions for participation in government contracts and more transparent procedures for the conferral of such contracts.

There are four basic principles regarding the placing of government contracts. First, there is a veto on technical specifications with a discriminatory effect. Second, government contracts must be advertised throughout the entire Union. Third, there must be objective criteria for participation. Fourth, there are various obligatory procedures in the placing of government service contracts.

The threshold value above which the Directive applies is 200,000 ECU for services and supplies and 5 million ECU for works.

The Co-ordination Directive applies to the state, to municipalities, water boards, public provisions and public institutes. European bidding rights are founded on Article 85 *et seq.* of the EEC Treaty. European law always takes precedence over national policy.

Eurocodes

Eurocodes form a separate category of European standards. They are standards intended for use in building structures, though their status is still unclear. In addition to those on general fundamentals, Eurocodes are in preparation in the fields of concrete, steel and reinforced concrete structures, timber structures, brick structures and foundations, earthquake-resistant structures, as well as for loads on structures. There were initially many views on the ways in which the Eurocodes could be operationalised. The most interesting of these suggests that the Eurocodes should be regarded as pre-standards, or optional EU directives, that, after necessary modifications, could function as European standards and as such would be given priority over the relevant national standards. Consultation between the Comité Européen de Normalisation (CEN) in Brussels and the European Commission has meanwhile led to the Eurocodes gaining the status of European standards.

After a transitional period of a few years the Eurocodes will replace national standards in the member states of the EU and the European Free Trade Association (EFTA).

European standards

In addition to the Eurocodes, the formulation of *European standards* is a matter of strategic importance. The Comité Européen de Normalisation (CEN) has been concerned with European standardisation since as early as 1961, not only in the building sector, but in most other sectors too. The members of CEN are the national standardisation institutes of those countries combined in the European Free Trade Association. The actual work of standardisation is performed by national standardisation institutes, in which experts from business and institutions are involved. The International Standardisation Organisation (ISO) is active on a world-wide scale. The CEN's activities have tended to be modest in the field of building, as in many others, first because of the absence so far of a direct relationship between the EU's harmonisation policy and the CEN's standardisation policy. Second, the methods so far employed by the CEN are largely inappropriate to the building sector, which hardly operates in national terms, let alone European ones. Third, not all countries are equally involved in the CEN's activities.

The near future, however, holds more favourable prospects for greater uniformity in regulations and standards, because of the close link that is now being established within the EU between policies of harmonisation and standardisation. Furthermore, since 1986 the CEN has changed its procedures. There is currently much greater emphasis on active programming by experts in various sectors and on project working. A new voting procedure has been adopted which replaces the need for unanimity – those member states in the minority are now also bound by the decision of the majority.

In 1993, the European Commission asked CEN-CENELEC to prepare a research mandate on the feasibility of the introduction of European standards for the qualification of construction companies. This was intended to lead to the harmonisation of the norms currently employed by the various member countries' own qualification and pre-qualification systems.

If these norms are implemented, it will mean that the guidelines on government contracts will be altered, referrals probably being inserted to obligatory European norms.

Harmonisation of declarations of quality and product liability

At the European level no agreement yet exists over definitions of product quality and a myriad of differing quality standards are in evidence. Hence the collective term 'declarations of quality' is used.

A good, workable scheme under which a European technical seal of approval can be obtained appears to be essential; it will enable new products to be marketed much more quickly. Because European standards still largely have to be developed, the conferral of technical approval is of great importance. In various member states, therefore, initiatives are currently being taken which should enable quality standards in one to be recognised in another. In both Great Britain and the Netherlands, for example, a Certification Board has been set up to regulate certification at a given quality level, and to counter proliferation within the certification institutions.

In July 1985 the Product Liability Directive was adopted by the EU Council. This directive relates to the amendment of legislative provisions of the member states concerning liability for defective products. For the construction and building supply industries this is not without importance.

It is now no longer necessary to prove that the producer is at fault; he or she is deemed automatically liable for damage caused by a defective product. In other words, the producer is at fault except under certain conditions laid out in the directive. However, the consequences for the building industry are limited. For though the directive is applicable to all movable property, and though this concept is interpreted so widely that building materials already incorporated into a building are also covered by the directive, the building as a whole is not itself a product within the meaning of the directive.

Harmonisation of tendering practices

Attempts to harmonise tendering practices started with the regulations affecting government contracts mentioned earlier. Whether more general regulations are to be introduced is not yet known. Regulations concerning the free movement of services between EU member states have been outlined in the case of government tendering in two directives. The first, the Liberalisation Directive, lays down (in Article 3) that member states are obliged to remove all barriers which inhibit the equal treatment of prospective contractors. The second, the Co-ordination Directive, aims at harmonising the legislative provisions of member states that are of direct application to the creation or operation of an internal market. This directive provides national legislators with guidelines for the consideration of tenders.

The Liberalisation Directive and the Co-ordination Directive should be viewed in relation to each other. It is well known that contractors often combine to form a price cartel, and that those seeking tenders in such cases are confronted in practice with a price that is the result of prior pricing agreements. In addition, present municipal tendering practice may be characterised as protectionist (Verjans 1987; Nijholt 1987, 1988). Foreign tenders, for instance, are disadvantaged, even in border areas. Nijholt has demonstrated this for a number of cases – Limburg, Liège, Aachen – where there proves to be hardly any sign of a transborder tendering policy. It was found that even in these border areas preference was given to local builders.

Under the EU rules for tendering, governments may not give preferential treatment to national or local firms when awarding contracts for building projects (Nijholt 1988). From 1990 all public works to be put out to tender which have a value of more than 5 million ECU must be reported to the European Commission in Brussels. Details of these projects are entered into a data bank, the *Tender Electronic Daily*, thus increasing the possibility of greater competition. It is also envisaged that all those tendering for a project should receive equal treatment. There is, however, a long way to go before real European tendering is established.

So far the Co-ordination Directive has proved a failure; hardly any government body has adhered to it. The fact that the new directives on government tendering continue to have little effect does not, however, mean that this will continue to be the case. In the case of large building projects in particular, there will be increasing pressure to give builders from different countries an opportunity to compete; and it is particularly in the case of large building projects that price and quality competition is most desirable.

The most immediate advantages of European economic unification will probably be felt by the larger building firms, particularly if they can adapt their market strategies and exploit the possibility of mergers and co-operative ventures with related enterprises in other EU countries. This is also a point which Bakens and Bergstein (1988: 27) have underlined: it will not be possible, they say, for smaller building firms to take on small and medium-sized projects in other countries. International competition in Europe will only be evident, for the time being at least, where large-scale projects are involved.

Throughout Europe, the size of individual companies in the construction industry varies widely. There are very large numbers of small building firms and a small number of large firms. Each size class has specialised in some part of the building market: small companies are devoted to maintenance, renovation, conversions and the smaller new buildings, while larger companies are better suited to the larger, more complex building projects. Small firms generally operate locally and regionally, the larger companies are more nationally and, in part, internationally oriented. It is only to be expected that smaller companies are likely to encounter little of the process of European integration, while larger companies will be confronted more directly with international competition. While this forms a threat, it also represents new opportunities.

Van Velzen (1988: 40), in his study of the effects of European integration, has concluded: 'Local, regional and national experts in design and implementation processes will not be troubled much by international competition' (see also Heerma 1990). This does not, however, tell the whole story. As a result of the boom that may well occur in building products and methods, building contractors and architects will be put on the defensive. Such a development should also be important for small and medium-sized contractors. If standards and regulations are harmonised at the European level, the internationally oriented building supply firms should be able to profit from the growth in the market more than the project-tied building contractors. Bakens and Bergstein (1988: 27) also point to the possibility that supply firms will in the future capture a substantial segment of the market from smaller contractors and subcontractors. Oskam, on the other hand, is of the opinion that there will be hardly any change in the level of imports and exports of building materials after 1992 (Oskam 1990).

Barriers for competition in the building industry

Building is necessarily tied to location, as a result of which the single market will in certain respects have little direct effect on construction. Product development in the building supply industry which is not tied to a specific project can, however, as a result of the development of a single market, exploit a much larger market (Reitsma and Bol 1991). We therefore expect the Europeanisation of the building sector to have a much greater effect on the building supply industry than on the construction industry, and, on balance, to strengthen the position of the former *vis-à-vis* that of the latter. Trade in building materials, building products and building systems will in general be furthered as a result of the integration of the European market.

It is interesting to observe that the new European directives on tendering for government contracts, in their aim to increase competition among building firms in member states and to encourage more tenders to be offered, are at variance with the pursuit of forms of public–private co-operation. This concept is open to many interpretations, not all of which are equally attractive. In general the government should be prepared to invite tenders from a number of contractors, along the lines outlined in EU directives. Where the government decides to proceed further with some of these tenders, the final contract should have to include provisions which make competition actual rather than simulated.

However, where public–private co-operation is advocated to encourage private sector financing of large building projects (such as toll tunnels), the question arises as to whether financing by government alone is not more effective. Where public–private co-operation is advocated to persuade the private sector (property development companies, etc.) to work together with the public sector, competition is in danger of being restricted, particularly if the development company and the construction firm form one concern. Co-operation between the public and the private sector is healthy, but only if it incorporates an element of competition so that the government works with the firm that has put in the most favourable tender.

In general, construction companies show a remarkable zeal for protecting themselves against competition. In theory, European integration should lead to healthier competition between construction companies, especially in larger building projects; in practice, the firm that carries out the work often employs all sorts of machinations to be able to protect itself from this competition.

For a start, language barriers still exist. How many Dutch construction firms could easily read and evaluate a job specification written in French? Then there are the lobbies. The construction and real estate sectors are typical contact-industries, where matters are often settled informally within networks that are able to resist the admission of certain participants. In many countries, representatives of the main contestants for a contract arrange deals in informal pre-negotiations; the official parts of the procedure look transparent enough, but behind the scenes the cards have been stacked long before the dealing begins.

Another particularly important hindrance to free competition is the matter of land ownership. Alert project developers and investors, moving faster and faster, have been buying up strategic sites which the government cannot subject to compulsory purchase orders as long as the owner is, in principle, prepared to implement the site's new function. At those locations where the Dutch government's 'Fourth Policy Document on Physical Planning Extra' specifies the planned building of new housing, large investors and project developers have already bought up every attractive site, establishing a practical monopoly. The Brussels regulations exert no effective grip in cases like this. Comparable situations exist in which contractors and capital market players have entered into informal partnerships, coalitions which then propose combined construction and financing deals. Finally, numerous international consortia are already seeking to strengthen their hold on parts of the European building market – and actually seek to subdue the competition (Root 1987).

All in all, it remains open to question whether the increased competition which European integration is said to have brought about will actually be felt in the European building market. There is, in any case, good reason to doubt it.

Different building processes

Full European integration of the construction market will be further complicated by the fact that the building process is still organised differently in different European countries. The Kolpron report 'The Construction Sector and Europe 1992' (Kolpron Management Support 1991) provides an overview of the structure of the building process in a number of European countries (see also Reitsma and Bol 1991). The report draws distinctions between project management, design management and construction management. The 'Project Manager' refers to the body actually commissioning the project, or the body appointed by them and entrusted with project responsibility and the authority to take project decisions. By 'design management' is meant responsibility for the processes of design, quantity surveying and drawing that precede construction. This is usually carried out by the architect, the builder, other engineers and so on. 'Building management' refers to the responsibility for the purchase of building materials and the construction itself. This can be illustrated by considering the different organisation of the construction process in four countries: Belgium, Germany, France and Great Britain.

Belgium

Around 80 per cent of building work follows one of two traditional patterns. In one case a construction firm is called in only after the definitive design has been agreed. The reason for this adherence to tradition is that architects have a monopoly position in the building process; there is a legal obligation to hire an architect for every building project for which a permit is required. The architect has a co-ordinating and supervisory role (see Figure 5.1). Alternatively, a project can also be entirely contracted out and in such cases the co-ordination is carried out by the main contractor (see Figure 5.2). In this second method (known in Britain as 'Design and Build') the client takes the project directly to a building contractor. It is found in industrial building and earthworks, roadbuilding and waterworks, as no architect need be called in for this type of work. In cases where

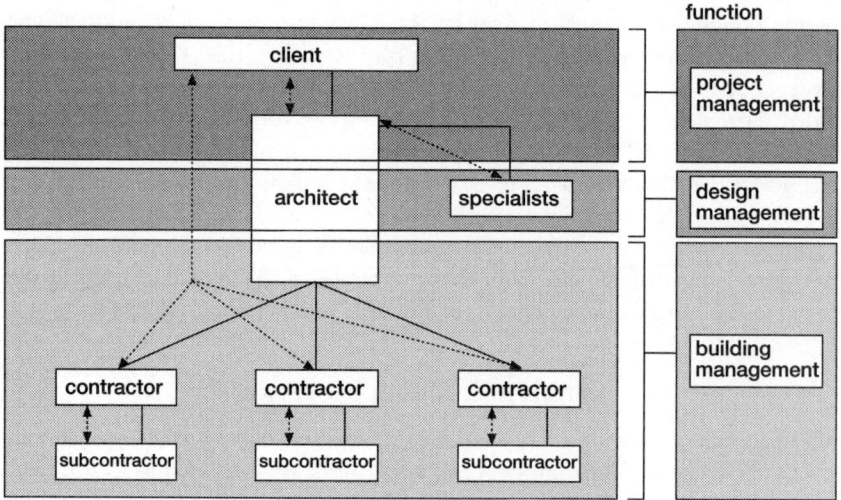

Figure 5.1 The traditional method in Belgium: the architect as co-ordinator

Source: Kolpron Management Support (1991).

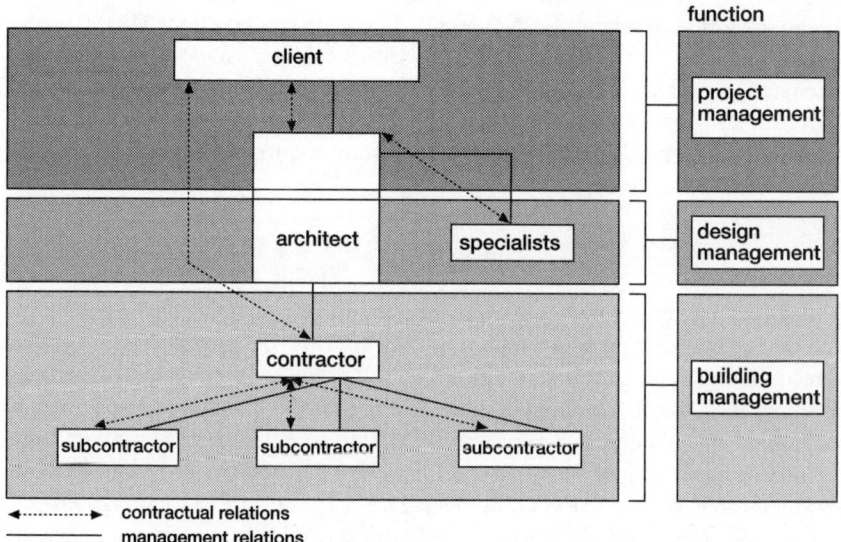

Figure 5.2 The traditional method in Belgium: the work is awarded to a main contractor

Source: Kolpron Management Support (1991).

the involvement of an architect is a legal requirement, the client is still entitled to prepare his own drawings and present them to the architect for endorsement.

Germany

In former West Germany, large projects are split up and the various parts awarded to various contractors, a process known as *Teillosevergabe*. In smaller building projects, the client hires an architect, a building company and an engineer. The architect is entitled to represent the client in contracts with the building company and the engineer (see Figure 5.3).

In more complex building projects, the client can enter three separate contracts: one with the architect, one with the engineers and one with the main contractor. Occasionally two architects are appointed, one responsible for the drawings and the other for the architectural site management. The main contractor is responsible for any subcontractors; while it usually carries out the structural work itself, much use is made of subcontractors for the woodwork, sanitary facilities and fittings, and so on (see Figure 5.4).

Architects' responsibilities and relevant fees are set out in the so-called *Honorarordnung für Architeckten und Ingenieure*. The appointment of an architect is a legal obligation in some federal states, but in others construction engineers are also entitled to apply for the building permit.

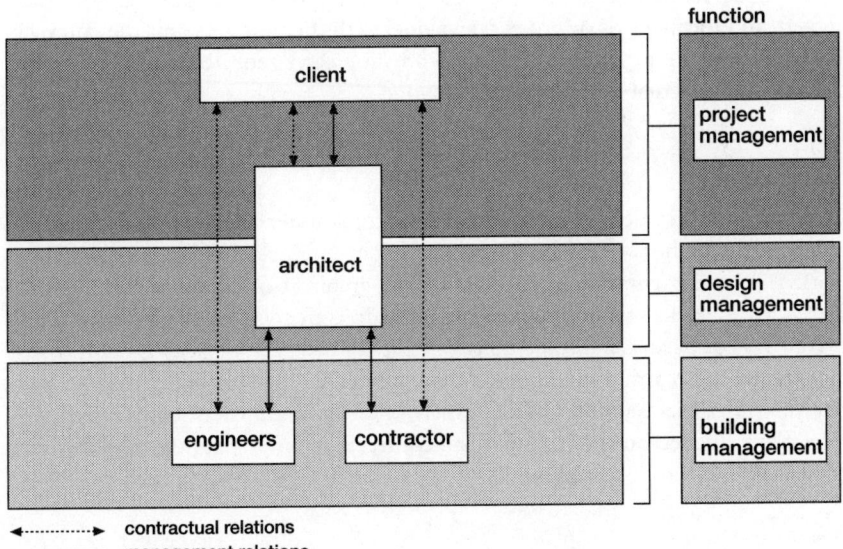

Figure 5.3 The building process for smaller projects in Germany
Source: Kolpron Management Support (1991).

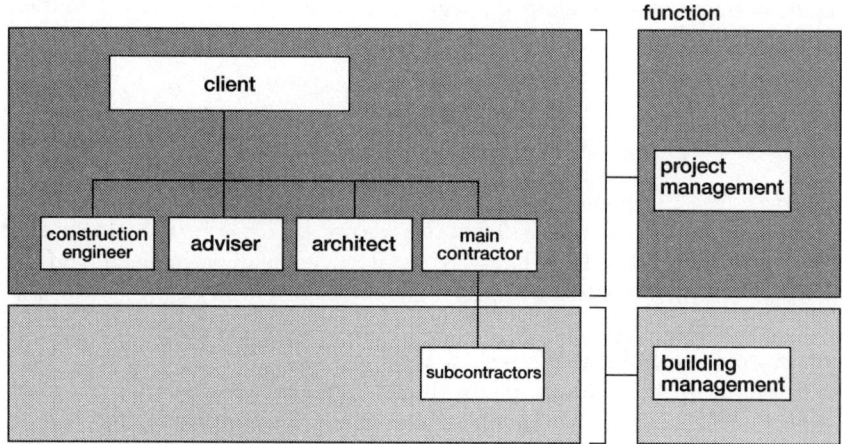

Figure 5.4 The building process for complex projects in Germany
Source: Kolpron Management Support (1991).

In Germany, as well as elsewhere, construction companies are increasingly adopting responsibility for the entire building project management, including the design stages. This is termed *Generalunternehmen*.

German public bodies (cities, municipalities, etc.) are obliged to keep to the terms of the *VOB* (the *Verdingungs-Ordnung für Bauleistungen*), which include open public tendering in all cases. In practice, 'investment bids' are publicly invited, and the investor or project developer with the best plan gets the contract. These investors and project managers are not then required to invite public tenders for the actual building contract work.

France

In France, project design has to be carried out under the responsibility of an independent architect (for buildings) or of an engineer (for civil engineering works). Private clients are entitled to hire an architect or a consultant engineer. The work is usually carried out in separate sections by a variety of contractors (see Figure 5.5) or by an established conglomerate of contractors (see Figure 5.6).

If the work is carried out by such a conglomerate, responsibility may lie with one or with all its members. This framework allows the client to exercise considerable influence on the choice of subcontractors. In France, the appointment of a chief contractor is unusual.

Great Britain

In Britain there are a number of ways of structuring the building process. First, there is the traditional method: the client appoints an architect, a builder, a

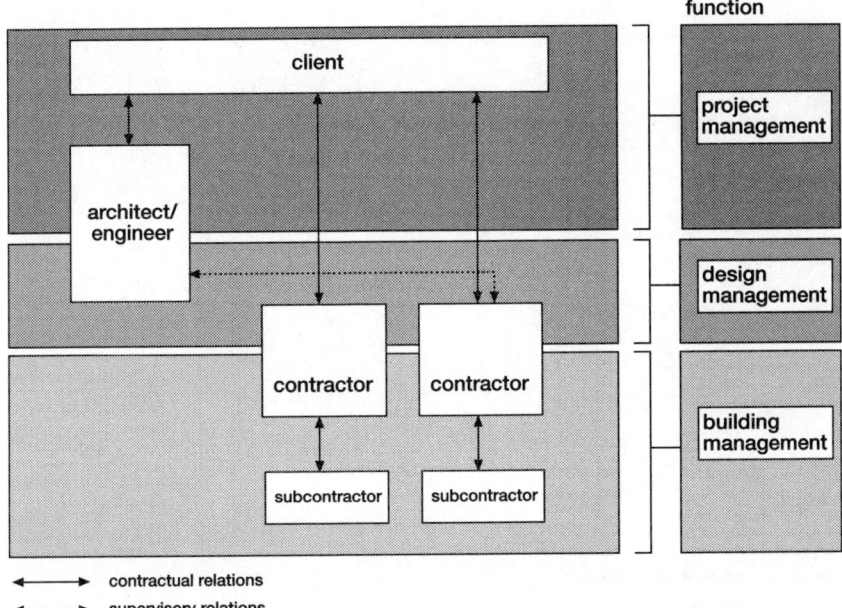

contractual relations
supervisory relations

Figure 5.5 French building work carried out by a number of contractors

Source: Kolpron Management Support (1991).

quantity surveyor, and often an engineer for the technical installations. These provide the design, the contract documentation and specifications, and the site supervision. The team is co-ordinated by the architect (see Figure 5.7).

This traditional method is becoming less popular. While over 75 per cent of building contracts (in financial terms) took place using this method in 1984, by 1989 this had dropped to 65 per cent.

Second, there is 'management contracting'. In this, a building contractor is brought into the design team, at a very early stage in the project, to provide management skills. This contractor replaces the architect as co-ordinator (see Figure 5.8). The method is applied in about 15 per cent of cases, a percentage which has remained reasonably stable over the years.

Third, there is 'construction management'. This method was introduced in 1989 and resembles management contracting. The difference is that the client enters into separate contracts with every subcontractor and pays them directly. The construction manager operates and manages independently and is even more important in the design process than was the case in 'management contracting'.

'Design and Build', an important alternative to the traditional method, is currently becoming a little less popular. In 1989 this method was employed in about 11 per cent of cases. This approach allows the client to locate accountability for both design and construction at a single point (see Figure 5.9).

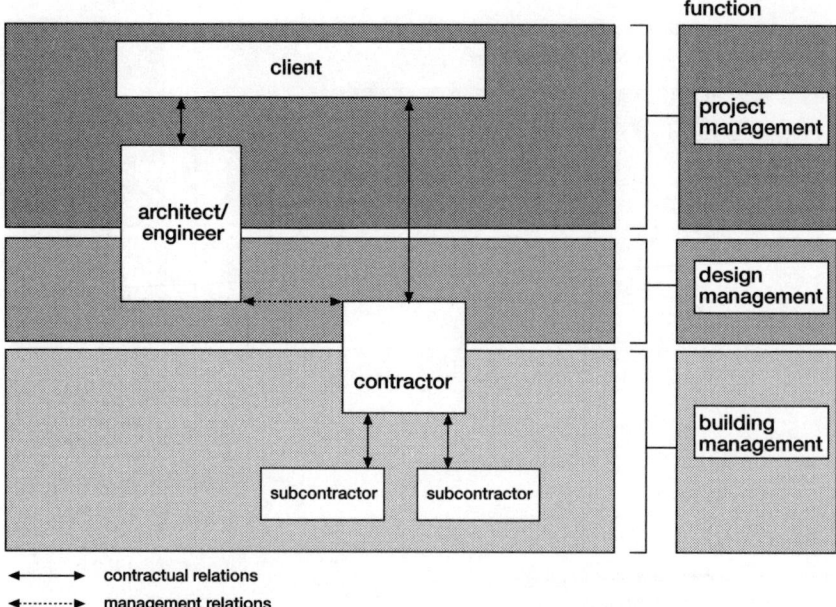

function

contractual relations
management relations

Figure 5.6 French building work carried out by a conglomerate of contractors
Source: Kolpron Management Support (1991).

If partners in a building project are accustomed to a given mechanism, they find it difficult to switch to another mechanism with a different distribution of responsibilities.

Conclusions

We can draw a number of tentative conclusions. First, the Europeanisation of the building industry is continuing. The harmonisation of technical regulations, standards, and guidelines concerning quality will affect the whole of the construction industry in the European Union. It is the larger building firms which will be initially and most directly affected by Europeanisation. It is also likely that there will be more mergers and joint ventures between partners in – and outside – the various EU member states. Second, the increase in market size will lead to a relative growth in the building supply industry compared with the construction industry. Third, the European directives on tendering are at variance with the efforts of most governments to encourage greater public–private co-operation. Fourth, as a result of the growing tendency to liberalise trade, there will be increasing pressure to abolish restrictive practices. It will make sense to trust in innovative capacity, keen pricing, and quality improvements as the principal weapons in the competitive struggle. Construction companies may be expected

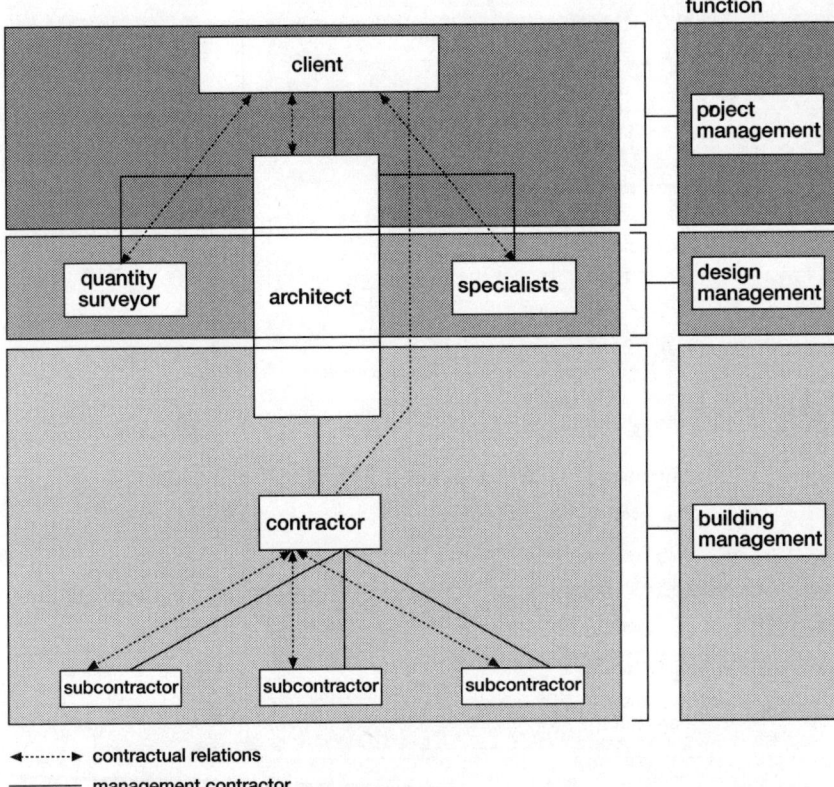

function

Figure 5.7 The traditional method in Great Britain

Source: Kolpron Management Support (1991).

to employ their enormous repertoire of competition-limiting practices to protect their positions in this market.

Despite some movement towards harmonisation, national differences in building processes are inhibiting the integration of the European building industry. The path to effective European tendering and a real European building market still appears to be a long one.

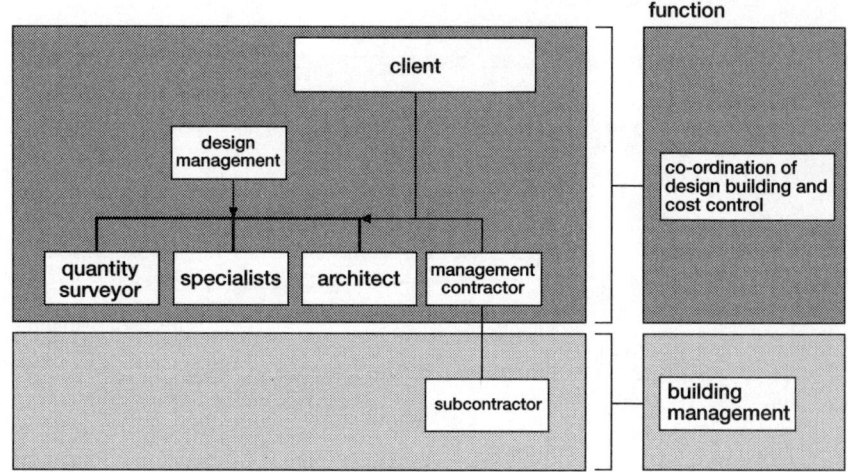

Figure 5.8 'Management contracting' in Great Britain

Source: Kolpron Management Support (1991).

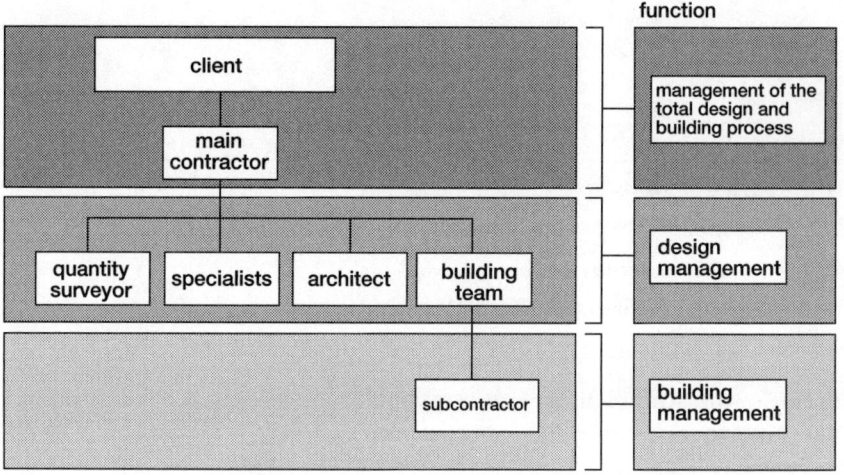

Figure 5.9 'Design and build' in Great Britain

Source: Kolpron Management Support (1991).

References

Bakens, W.J.P. and Bergstein, L.P.A. (1988) 'Overtrokken voorstelling van mogelijk-heden buiten Europa' (Exaggerated Picture of Possibilities outside Europe), *BOUW* 43 (14 Oct.): 26–29.

Heerma, E. (1990) *Nederlandse volkshuisvesting en Europese samenwerking en integratie* (Dutch Housing and European Co-operation and Integration), Tweede Kamer 1989–1990, 21.623 no. 1, The Hague.

Kolpron Management Support (1991) *De bouwsector en Europa 1992* (The Building Sector and Europe 1992), Rotterdam (SBR) (5 vols.).

Langeveld, A. and de Vries, D. (1990) 'Op naar een Europese bouwregelgeving' (Towards a European Building Regulation), *Corporatie Magazine* 1, 4: 21–23.

Nijholt, H. (1987) *Aanbestedingsrecht en aanbestedingspraktijk in de Nederlandse Euregio* (Tendering Law and Tendering Practices in the Dutch Euregio). Maastricht: Maastricht University Press.

Nijholt, H. (1988) 'EG-regels voor aanbestedingen brengen gemeenten in problemen' (EG-regulation for Tendering Causes Problems for Municipalities), *Volkskrant* 22 (Oct.).

Oskam, E.A. (1990) *De internationale handel in bouwmaterialen in het licht van het Europese integratieproces* (The international building materials trade in the perspective of the European integration), Amsterdam: EIB.

Pelkmans, J. (1985) *Opheffing van technische handelsbelemmeringen in de EG* (Elimination of technical trade barriers in the EC). The Hague: VNO.

Priemus, H. (1991) 'Unification of the European Building Market: Possible Consequences for the Dutch Construction Industry', *Netherlands Journal of Housing and the Built Environment*, 6, 1: 35–45.

Reitsma, D. and Bol, N.J.T. (1991) *Bouwtoeleverende industrie in Europa 1992: een oriëntatie voor België, Duitsland, Frankrijk en Groot-Brittannië* (Building supply industry in Europe 1992; an orientation for Belgium, Germany, France and Great Britain), Rotterdam (Stichting Bouwreseach).

Root, F.R. (1987) *Entry Strategies for International Markets*. Lexington, MA: Lexington Books.

van Velzen, N. (1988) 'De integratie van de Europese markt' (The integration of the European market), *Woningraad* 19: 36–40.

Vereniging van Boorondernemers en Buizenleggers, Nederlandse Vereniging van Wegenbouwers, Vereniging Aannemers Grond-, Water- en Wegenbouw (1991) *Europa 1992, Grond-, water- en wegenbouw* (Europe 1992, land, water and road building), The Hague: VBB, NVW and VAGWW.

Verjans, J. (1987) *Aanbestedingsrecht en aanbestedingspraktijk in de Belgische Euregio* (Tendering law and tendering practices in the Belgian Euregio), Maastricht: Maastricht University Press.

Walters, W. (1988) 'Europese bouwmarkt op normen gebaseerd' (European building market based on norms), *BOUW* 43 (14 Oct.): 66–67.

Part II

POLICIES

6

FISCAL RESTRAINT AND HOUSING POLICIES UNDER ECONOMIC AND MONETARY UNION

Mark Stephens

Whilst it was always obvious that a European single currency would have significant implications for the conduct of monetary policy, the impacts of Economic and Monetary Union on fiscal policy were much less widely appreciated, until governments began to cut budgets to qualify for currency membership. This chapter addresses the relationship between Economic and Monetary Union and the costs of member states' housing policies. The second section outlines the European Commission's economic rationale for constraining budget deficits within a monetary union, and the third outlines the institutional mechanisms for achieving this goal. The fourth section examines the current fiscal position of the member states and the fifth examines strategies which have been adopted, or may be adopted, to limit public expenditure on housing. Conclusions are drawn in the sixth and final section.

The economics of fiscal restraint

The Treaty on European Union (TEU) commits the member states and the Community to 'compliance with the following guiding principles: stable prices, sound public finances and monetary conditions and a sustainable balance of payments' (TEU, Article 3a(3)). Such principles imply a firm rejection of previously commonly held economic objectives, such as full employment and economic growth. The key institutional development in the TEU is of course the introduction of a single currency and therefore monetary policy managed by the independent European Central Bank, which, when established, will be bound to pursue price stability by the Treaty (Article 105).

There are several possible justifications for a single currency. The basis of two such justifications is that they will remove key barriers to the creation of the

European single market: the removal of currency risk and uncertainty from trade, borrowing and any other transactions previously employing different currencies within the EU, and the reduction in transaction costs incurred when changing currencies or undertaking currency swaps to minimise currency risk. Indeed, this is implied by the title of the European Commission's lengthy economic justification of the proposed single currency, 'One Market, One Money' (CEC 1990). However, the Commission places much greater emphasis on the general economic case for a single currency, which is based on the benefits of price stability.

Clearly, any shift towards a single currency implies the removal of monetary policy (interest and exchange rate decisions) from the governments of member states. This is justified because activist monetary policies are perceived to be self-defeating. Governments that devalue their currencies will experience no long-term gains to 'real' economic variables (such as output, investment and employment) because domestic prices will rise. The real value of economic growth will be eroded by higher prices and these, in turn, will remove the competitive advantage gained by devaluation.[1] If the optimal monetary policy is one dedicated to price stability, then one may as well share it with other countries and gain the benefits of currency stability.

However, a single currency also has implications for fiscal policy. There are two principal concerns expressed by the Commission. First, there is the fear that member states might accumulate unsustainable budget deficits. When a country runs up an unsustainable deficit, two possibilities arise. Either the government can 'monetise' the debt (that is its central bank will issue cash) which is inflationary, or it will default (CEC 1990: 106). Clearly, monetisation is not an available option for governments without currencies, although they could still default. The Commission examines the options:

1   The European Central Bank (ECB) could soften its monetary stance to lower the member state's loan repayments. This could avert a financial crisis. However, central bank independence and its anti-inflationary credibility could be undermined. This could push up interest rates.
2   The ECB could take no action, but the member state might leave the monetary union in order to monetise the debt. Again a key element of anti-inflationary credibility, in this case the irrevocable nature of the monetary union, would be undermined. Markets could respond with higher interest rates.
3   The ECB could bail out the member state through the purchase of a disproportionate share of public bonds from a specific country, but this would affect the overall market for government paper (CEC 1990: 107).

No matter how excessive deficits are tackled they will always have a deleterious impact on the monetary union as a whole. The Commission therefore identifies a need for 'fiscal discipline' within a monetary union, which is defined as the avoidance of the build up of unsustainable deficits (CEC 1990: 107).

But this is not the end of the matter. The Commission argues that even sustainable deficits can be too high, citing the example of their impact on the common dollar exchange rate (CEC 1990: 107). It is also argued that the impacts of 'crowding out' (rising interest rates caused by government borrowing) would be diluted through monetary union because the pool of (same currency) funds would be larger (CEC 1990: 113). It is nevertheless conceded that in the control of sustainable deficits, there may be a conflict of interest between the country and the Community as a whole (CEC 1990: 107), but this might well be a free-rider problem. The Commission identifies the need for 'regular policy coordination and surveillance' to avoid undesirable, but sustainable budget deficits (CEC 1990: 108).

The Commission does not, however, argue for 'fiscal federalism', in the form of a greatly enlarged EU budget, which might be justified on several grounds. The spillover effects of domestic fiscal policies might be such as to justify day-to-day interference in domestic fiscal policies, an argument the Commission rejects for empirical reasons (CEC 1990: 100). Another argument would be that the loss of monetary policy instrument implies a greater role for fiscal policy in the event of destabilising economic shocks. The MacDougall Report (CEC 1977) noted that federal systems were invariably based on a redistributive federal expenditure accounting for some 25 per cent of GDP, compared to 1 per cent (now 1.25 per cent) in the European Community. The Commission assumes that a budget of this size will not be available, while it expresses a preference for adjustments to be made through wage–price flexibility rather than through fiscal activism (CEC 1990: 101–102). Others argue that such stabilisation can be conducted at the level of the member states. The need for an activist fiscal policy should not be confused with the need for a centralised one (Allsopp et al. 1995: 141), and the Commission does concede that 'fiscal policies can alleviate temporary country specific disequilibria . . . [and] Budgetary adjustments can also be a necessary medium-term component of the path towards a new equilibrium in the case of permanent shocks . . .' (CEC 1990: 102). This does not imply a need for a centralised budget, and neither does the response to a Community-wide shock presumably because coordinated fiscal action could be arranged. However, the Commission's term 'fiscal autonomy' seems too strong to describe the acceptability of variations in member states' fiscal deficits 'in the medium run' (CEC 1990: 106).

The emphasis of the Commission's argument is overwhelmingly on the need to control deficits. Deficits can, of course, be controlled by raising taxes as well as limiting expenditure. Nevertheless, there are limits to the extent to which a government can raise taxation: aside from reasons of electoral popularity and the impact on incentives, in a single market taxable factors are far more inclined to migrate to jurisdictions with lower tax rates. This might be true of the corporate sector in particular and even to some segments of the labour market. The Commission does acknowledge that there is a need to maintain certain levels of expenditure in order to maintain the adequate provision of public goods. The

Commission recommends that 'minimal standards and common rules should be set when necessary' (CEC 1990: 101). It is notable, however, that it is public goods that are referred to: that is goods (such as defence) which must be provided by the state because their consumption cannot be restricted. Public goods are not defined as goods or services provided for the public or by the state; plainly, social housing is not a public good.

Mechanisms for restraining budget deficits

The TEU reflects the two primary fiscal concerns of the European Commission, i.e. excessive deficits and multilateral surveillance. There are three discernible mechanisms through which the Commission and Council exert influence over member states' budget deficits.

Convergence

The strongest sanction that the EU has over member states' budgets arises from the requirement for them to meet four convergence criteria (laid out in a protocol to the Treaty) before they can join the monetary union (assuming, of course, that they wish to do so). The four convergence criteria are:

1 inflation must be no higher than 1.5 percentage points of the three lowest rates in the year before examination;
2 at the time of the examination, a member state must not be subject to a Council decision that an excessive deficit exists;
3 a member state must have respected the normal fluctuation margins provided within the ERM without severe tensions for two years before the examination, and in particular must not have devalued unilaterally during the same period; and
4 the average nominal long-term interest rate over the year up to examination must not be more than 2 percentage points higher than the average of the three countries with the lowest inflation rates (TEU, Convergence Criteria, Articles 1–4).

   In pursuit of economic convergence, member states have been required to adopt 'where necessary . . . multiannual programmes intended to ensure the lasting convergence necessary for the achievement of economic and monetary union, in particular with regard to price stability and sound public finances' (TEU, Article 109e(2)). In practice, all member states other than Luxembourg submitted convergence plans (von Hagen and Harden 1994). The only tangible sanctions regarding convergence programmes relate to the four 'cohesion states'. Ireland, Spain, Portugal and Greece each had per capita GDPs less than 90 per cent of the Community average and consequently qualified for assistance from the Cohesion Fund, established by the TEU. The Fund may be used for 'projects in the fields of environment and trans-European networks', but countries must 'have a programme leading to fulfilment of the conditions of economic convergence' (TEU, Protocol on Economic and Social Cohesion). In practice the convergence programmes under Article 109e(2) have served this function (von Hagen and Harden 1994).

Excessive deficits

One of these convergence criteria is that a member state 'is not subject of a Council decision ... that an excessive deficit exists' (TEU, Convergence criteria, Article 2) which the main Treaty also commits member states to 'endeavour to avoid' in the run up to monetary union. To this end, the Commission is charged with monitoring governments' budgetary positions and their stock of outstanding debt. Budgetary discipline is assessed on the basis of the debt to GDP ratio and the ratio of government debt to GDP according to two specified 'reference values'. The reference value for the debt to GDP ratio is 3 per cent and 60 per cent for the ratio of the stock of government debt to GDP (TEU, Article 104c(2) and Excessive Deficit Procedure, Article 1). In both cases the Commission must also consider the direction of the deficits and, in the former's case, whether any excess is only 'exceptional and temporary' (TEU, Article 104c(2)). Should a member state fail to meet either or both of these criteria, the Commission must prepare a report, which also takes into account 'all other relevant factors' (TEU, Article 104c(3)). Following consultation with the Monetary Committee, the Commission then decides whether it considers there to be an excessive deficit. Only if the Commission believes there to be (or that there may be) an excessive deficit does the Council (in the form of Economics and Finance Ministers (Ecofin) using qualified majority voting) get the opportunity to decide whether one does, in fact, exist 'after an overall assessment' (TEU, Article 104c(6)). As von Hagen and Harden (1994) have observed, the Commission has a powerful 'gatekeeper' function in that the Council cannot decide whether an excessive deficit exists without the Commission first deciding that there is one, and both the Commission and the Council have a great deal of discretion when making such an assessment.

The avoidance of an excessive deficit decision by the Council is not the only sanction against member states which fail to operate sufficiently prudent fiscal policies. From 1994, when a member state is subject to an excessive deficits decision, the Council is obliged to make recommendations to that member state with a view to removing the excessive deficit within a specified period. Recommendations are not legally binding, but should the member state fail to take 'effective action', the Council can make its recommendations public (TEU, Article 104c(8)).

The commitment to fiscal restraint remains after a country has joined the monetary union, and it is at Stage 3 (the point at which exchange rates are fixed irrevocably) that sanctions become more tangible (TEU, Articles 109e(4) and

104c(1)). Should the member state persistently ignore the Council's recommendations, the Council has discretion to notify the member state of measures for deficit reduction that it deems to be necessary, and it may request that the member state submits reports according to a timetable (TEU, Article 104c(9)). Following failure to comply at this stage, the Council is then permitted to operate actual sanctions against the offending member state. The Council may:

1   require the member state to publish additional information before issuing bonds and securities (in effect a health warning is attached to government debt);
2   'invite the European Investment Bank to reconsider its lending policy towards the member state concerned'; and
3   require the member state to make a non-interest-bearing deposit ('of an appropriate size') with the Community until the Council is satisfied that the excessive deficit has been corrected;
4   impose fines 'of an appropriate size' (TEU, Article 104c(11)).

The principle of imposing fines is a key part of the 'stability pact' agreed at the European Council in December 1996.

Multilateral surveillance

In line with the Commission's view that even sustainable deficits may have undesirable externalities, the TEU put in place procedures for the 'multilateral surveillance' of member states' economic policies, starting in 1994, and applying to all member states regardless of whether they are members of the monetary union.

Through a convoluted process, the Council is charged with drawing up recommendations for the 'broad guidelines' for the conduct of economic policy by the member states and the Community. *En route*, the Commission makes recommendations to the Council and the Council's final decision (made under qualified majority voting) is based on a conclusion of the European Council (TEU, Article 103(2)). Member states are obliged to provide information to the Commission regarding economic policy. The Commission then reports to the Council and on this basis monitors the economic policies of member states in connection with the 'broad guidelines' (TEU, Article 103(3)). Should the Commission decide that a member state has failed to follow the 'broad guidelines' or that a member state's policies 'risk jeopardising the proper functioning of economic and monetary union', then the Council can make recommendations to the member state, which it may choose to make public (TEU, Article 103(4)). Recommendations are not binding, so the multilateral surveillance provisions do not appear to be backed with such tangible sanctions as the excessive deficits procedure. However, since fiscal deficits are still covered by the excessive deficits procedure and monetary policy by the European Central Bank, the vagueness of multilateral surveillance does not in itself offer much scope for divergent economic policies.

The fiscal position of the member states

The fiscal position of the member states over the period 1990–1995, with the expected outcome for 1996 and the forecast for 1997, is summarised in Tables 6.1 and 6.2. The outcome for 1997 will be used as the basis for deciding which member states qualify for membership of the single currency. The tables are based on the two 'reference values' that contribute to the assessment as to whether a member state has an excessive deficit: the general government deficit and the stock of general government gross debt, both expressed as percentages of GDP. The 'reference values' in the Treaty are 3 and 60 per cent respectively.

General government borrowing

The level of general government borrowing has tended to rise since 1990. This deterioration in part reflects the impact of the recession which has the effect of increasing expenditure on items such as social security benefits, while simultaneously reducing tax revenues. But, even in 1990, only five member states met the 'reference value' of 3 per cent (compared to two in 1993, the worst year). There is of course much variation between countries, but in late 1996, seven countries were forecast to meet the 'reference value' in the crucial year of 1997. These included some 'close calls' and in early 1997 there was concern about the extent to which rising unemployment would affect Germany's ability to qualify.

Given that the European Monetary Institute (EMI) noted that the deficit in 1993 (6 per cent of Community GDP) was the highest recorded since the creation of the European Economic Community and that improvements in 1994 were cyclical rather than structural (EMI 1995: 2), the forecast implies actual or planned attacks on structural deficits. For example, Sweden is expected have cut its deficit from 13.3 per cent in 1993 to just 3.1 per cent in 1997.

General government gross debt

In 1994 only four countries met the reference value relating to outstanding debt. Further, even those countries which remain within the reference value generally experienced large rises in the first half of the decade, and these are expected to continue to rise (e.g. Germany, France and the UK). Some countries have such large outstanding debts it is clearly not possible for them to reach the reference value in the foreseeable future: Belgium, Greece and Italy stand out in this respect. Ireland is exceptional in managing to contain and reduce its deficit. Each of the new EU members had outstanding deficits above the reference value in 1994, although Austria's was the lowest. Only the UK, France and Luxembourg

Table 6.1 General government lending (+)/borrowing (−) (as per cent of GDP)

Country	1990	1991	1992	1993	1994	1995	1996	1997
Belgium	−5.4	−6.5	−6.7	−6.6	−5.5	−4.5	−3.2	−3.7
Denmark	−4.5	−2.1	−2.5	−4.4	−4.3	−1.4	−0.9	−0.6
Germany	−2.1	−3.2	−2.6	−3.3	−2.9	−3.5	−3.9	−2.9
Greece	−14.0	−13.0	−11.7	−13.3	−14.1	−9.2	−8.1	−6.9
Spain	−3.9	−4.9	−4.2	−7.5	−7.0	−6.2	−4.8	−3.7
France	−1.6	−2.2	−3.9	−5.8	−5.6	−5.0	−4.2	−3.0
Ireland	−2.2	−2.1	−2.2	−2.5	−2.4	−2.4	−2.0	−1.6
Italy	−10.9	−10.2	−9.5	−9.5	−9.6	−7.1	−6.3	−5.2
Luxembourg	+5.9	+2.3	+0.3	+1.1	+1.3	+0.3	−0.7	−0.3
Netherlands	−5.1	−2.9	−3.9	−3.3	−3.8	−3.4	−3.5	−2.9
Portugal	−5.5	−6.6	−3.3	−7.2	−6.2	−5.4	−4.4	−3.7
UK	−1.5	−2.6	−6.2	−7.8	−6.3	−6.0	−4.4	−3.7
EU-12	−4.0	−4.6	−5.0	−6.0	−5.6	...	...	...
Austria	...	...	...	−4.1	−4.4	−6.2	−4.6	−3.1
Finland	...	...	...	−7.2	−4.7	−5.6	−3.3	−1.6
Sweden	...	...	...	−13.3	−11.7	−8.1	−5.2	−3.1
EU-15	...	...	...	...	...	−5.0	−4.4	−3.4

Source: 1990–1994: European Monetary Institute (1995: Tables 4 and 8, and Box 3); 1995–1997: European Commission.

Notes
1996 figure expected out-turn; 1997 forecast.

Table 6.2 General government gross debt (as per cent of GDP)

Country	1990	1991	1992	1993	1994	1995	1996	1997
Belgium	130.8	132.9	133.8	138.9	140.1	133.7	132.2	130.6
Denmark	59.6	64.6	68.8	79.5	78.0	71.9	71.0	68.7
Germany	43.8	42.1	44.8	48.1	51.0	58.1	61.5	62.4
Greece	82.6	86.1	92.3	115.2	121.3	115.5	111.8	111.4
Spain	45.1	45.9	48.2	59.8	63.5	65.7	67.8	68.0
France	35.4	35.8	39.6	45.8	50.4	52.4	56.1	57.8
Ireland	96.8	96.2	93.4	96.1	89.0	85.5	81.3	77.3
Italy	97.9	101.3	108.4	118.6	123.7	124.8	124.5	122.8
Luxembourg	5.4	4.9	6.0	7.8	9.2	5.9	6.2	6.8
Netherlands	78.8	78.9	79.9	81.4	78.8	79.0	79.4	78.7
Portugal	67.7	69.3	61.7	66.9	70.4	71.6	72.2	71.8
UK	...	35.8	42.0	48.3	50.4	54.0	55.5	56.2
EU-12	...	57.0	60.8	66.1	68.9	...	...	...
Austria	...	...	...	63.5	65.0	69.4	72.4	73.9
Finland	...	...	...	61.8	70.0	59.6	62.5	63.2
Sweden	...	...	...	74.4	81.0	79.9	80.8	79.6
EU-15	...	...	...	...	...	71.2	73.9	74.3

Source: 1990–1994: European Monetary Institute (1995: Tables 4 and 8, and Box 3); 1995–1997: European Commission.

Notes
1996 figure expected out-turn; 1997 forecast.

are expected to meet this reference value in 1997, but as described above, the reference value is just that; it is not a hard and fast rule.

Overall assessment

The EMI has noted that 'at present most EU countries would not qualify for Monetary Union, in the majority of cases because of their fiscal positions' (EMI 1995: 5). Indeed, in the first year of its operation, all member states received a Council decision to the effect that they had excessive deficits, with the exceptions of Ireland and Luxembourg. (In the case of Ireland, the improvement in the outstanding debt in 1994 was judged to be sufficient to avoid an excessive deficit, even though it remained above the reference value.)

The EMI claims that the deterioration in deficits in the 1990s was due to a continuation of a long-term weakening of structural fiscal positions, and not just the recession. However, it attributes the improvement in 1994 to purely cyclical factors (EMI 1995: 53–54). The EMI concluded that the economic recovery should be used for 'undertaking sweeping reforms of the function and size of the public sector' (EMI 1995: 54).

Housing and public sector deficits

In the mid-1990s, most member states faced relatively large budget deficits and they have made efforts to reduce them. The desire to join the single currency has been a stated motivating factor in virtually all cases, although there have been other pressures too, such as a desire to cut the level of taxation and the need to bring the cost of pension systems under control. When governments take steps to reduce budget deficits it will not always be possible to discern the primary motivation. For example, a government may use the reference values in the TEU as a reason to make unpopular expenditure cuts to give the impression that the decision is outside its control. But the government may be keen to cut the deficit anyway. While the TEU is not the only source of budget pressures, it is an important one, and with entry into the single currency being determined on the basis of the 1997 deficits, it has added urgency to expenditure cuts.

While decisions on monetary union will be made using standard measurement, not all of the member states use the same definitions of public expenditure as the primary indicator of their fiscal position for domestic economic policy. The UK uses the widest definition of public expenditure which covers all of the public sector including the financial deficit of public corporations, whereas most member states use expenditure criteria which exclude the 'own account' borrowing of public corporations (Hawksworth and Wilcox 1995: 25). The TEU specifies that the general government definition should be used for the purpose of assessing member states' fiscal positions. This includes 'central government, regional or local government and social security funds, to the exclusion of commercial operations' (TEU, Excessive deficit procedure, Article 2). It is therefore most at variance with the measurement used by the UK. Of course expenditure is only one part of the deficit equation: tax reliefs are not normally treated as expenditure, but the consequent foregone revenue has exactly the same effect on a government's deficit as a formal expenditure programme.

There is no standardised definition of public expenditure on housing or systematic collection of statistics. Those statistics which have been collected are therefore not strictly comparable and should be treated with caution. Further, it is important to consider the context in which a country's housing policy operates. For example, a high expenditure on housing allowances may be indicative of low insurance benefits rather than a generous housing policy (McCrone and Stephens 1995: 3). Measurements of expenditure on housing also include the cost of tax reliefs, and are generally expressed as a proportion of national income to allow for cross-national comparison.

Table 6.3 compares the most recently available cost of housing policy with the size of the general government deficit in 1994. The figures suggest that there is no simple link between the cost of housing policy and the size of the budget deficit. For example, Belgium and Spain appear to spend relatively little on housing policy, but each has a large budget deficit. This suggests that in these countries savings in housing policy cannot contribute significantly to reducing the budget deficit. Conversely, the Netherlands devotes quite a high proportion of its GDP to housing policy, but it also has one of the lower budget deficits. However, in the earlier part of the 1990s, Sweden and the UK both devoted high proportions of their GDPs to housing policy and had large budget deficits. At the time this may have suggested that in Sweden and the UK reductions in the costs of housing policy could play a major role in deficit reduction strategies.

Table 6.3 The costs of housing policy and government deficits (as per cent GDP)

Country	Cost of housing policy (year)	General government deficit (1994)
Belgium	0.24 (1988)	5.5
Denmark	1.02 (1988)	4.3
Germany	1.40 (1991)	2.9
Spain	0.98 (1990)	7.0
France	1.80 (1993)	5.6
Netherlands	3.20 (1990)	3.8
UK	3.30 (1992–3)	6.3
Finland	1.48 (1987)	4.7
Sweden	4.10 (1992)	11.7

Sources: Papa (1992: Table 10.8) except Germany, France, Netherlands, UK and Sweden (McCrone and Stephens 1995), Spain (MOPT 1992) and Finland (Hedman 1994: 131).

But, as has been suggested, it is not as simple as this, and one would need to examine the nature of housing expenditure in these countries before making judgements. Table 6.4 attempts to break down the main components of housing policy costs in six countries into three categories: tax reliefs, housing allowances and direct subsidies. Tax reliefs to owner-occupiers are taken to mean mortgage interest tax relief, except in Germany where the depreciation allowance is the primary tax relief for owner-occupiers. There is no attempt to measure other fiscal costs, such as the absence of a tax on imputed rent (which is indeed not taxed to any significant extent in any of these countries other than the Netherlands), or the virtual non-taxation of housing capital gains (effectively absent in these countries, including Sweden where roll-over relief was introduced). The table suggests that tax reliefs to owner-occupiers are most significant as a housing subsidy in Spain, which is unsurprising given the absence of housing allowances and the large size of the owner-occupied sector. However, it is also available on very generous terms in Spain and there have been no attempts to target it more effectively. The generosity of mortgage interest tax relief in the Netherlands is somewhat misleading because of the taxation of imputed rent. In the UK, the importance of tax relief has diminished greatly in recent years (it recently accounted for 43 per cent of the total cost of housing policy), partly due to reductions in the rate at which it can be claimed, but also because interest rates have fallen. Sweden and Finland have also made reductions in mortgage interest tax relief: in Sweden the proportion of interest costs eligible for relief has been reduced while in Finland the relief now applies to tax liability, so is no longer worth more to higher rate tax payers. In contrast, the cost of Germany's depreciation allowance has risen. Although time limited (to eight years) it can be claimed at the owner's marginal rate of tax. In some countries, therefore, there would appear to be scope for further reductions in mortgage interest tax relief.

Several factors account for the varying importance of housing allowances. They may of course be relatively important if they are generous, but this may not

Table 6.4 Components of housing policy costs (as percentage of total costs)

Country	Tax reliefs to owner-occupiers	Housing allowances	Direct subsidies
Spain	47.5	0	43.1
France	8.9	47.0	12.2
Germany	27.2	10.3	27.7
Netherlands	34.1	10.8	53.4
Sweden	24.9	23.1	52.1
UK	26.1	40.0	16.1

Sources: Papa (1992: Table 10.8), except Germany, France, Netherlands, UK and Sweden (McCrone and Stephens 1995), Spain (MOPT 1992) and Finland (Hedman 1994: 131).

Note
Figures do not necessarily sum to 100 since not all housing expenditure falls into these categories.

be the most likely explanation. They may also reflect the lack of generosity of other social security benefits and are likely to be sensitive to factors such as rising unemployment. In this respect, they are a demand-led expenditure and as such more difficult to cut. They can also reflect the extent to which governments have switched assistance away from general direct subsidies and towards selective assistance towards the poorest households, a common policy trend in the 1980s.

Direct subsidies take a variety of forms. In Spain, they mainly consist of interest subsidies to individual owner-occupiers, although subsidies are more generous for low income families and for first-time buyers. In Sweden they reflect interest subsidies on loans in each of the tenures. The open-ended nature of these interest subsidies caused the government to introduce a major structural reform in 1993 to limit the state's liability. In the UK, direct revenue subsidies to local authorities to subsidise rent levels were greatly reduced in the 1980s, and the figure includes the capital subsidies to housing association development programmes. Reductions in direct subsidies with increased dependence on selective assistance through housing allowances may appear a wholly justified strategy when budgets are constrained. However, such policy shifts have been associated with 'ghettoising' parts of the social rented sector as it becomes less attractive to better-off tenants. Increased reliance on means-tested housing allowances may also reduce work incentives as benefit is withdrawn as income rises. Such concerns appear to be behind the German government's decision to vary direct subsidies for new social rented housing developments according to the general expected income levels of the future tenants. This marks a significant reversal in the general trend in housing policy in Europe over the past decade (Maclennan 1995: 15).

The ownership of social housing may also have some bearing on the costs of housing policy. This is illustrated in Table 6.5, which includes the member states with the largest stocks of social rented housing. In each case, other than the UK, borrowing from the private sector by social housing companies is not treated as government expenditure, under either the definitions of expenditure used in those countries, or the definition used for the debt reference values. Even in

Table 6.5 Accounting and financial control of social housing sector

Country	Primary agency	Sectoral classification	Key budget deficit measures
France	HLM	Private and public corporations	Central government
Germany	Non-profit companies	Private and public corporations	General government
Netherlands	Housing associations	Private corporations	Central/general government
Sweden	Local housing companies	Public corporations	Central/general government
UK	Council housing	Local government	Public sector

Source: Hawksworth and Wilcox (1995: Table 3.1).

Sweden, where most social rented housing is owned by local housing companies, these are unambiguously in the public corporate sector and borrowing by them does not score as public expenditure, although subsidies from government obviously do. The difficulty faced by local authorities as landlords in the UK is that their housing is undeniably part of the government sector so even unsubsidised borrowing from the private sector scores as public expenditure under both the UK definition and that used by the European Commission. This is one reason why local authority investment in housing has been so restricted by central government in the UK, and why non-profit sector housing associations have been chosen as the main vehicles for new social rented housing in the UK since 1988. A recent report suggested that if the UK were to adopt the definition of public expenditure employed by the European Commission (that of general government expenditure), then local authority housing could escape from these public sector restrictions, provided that they could be restructured as 'quasi corporation', that is they:

- charged economically significant prices;
- operated and were managed in a similar way to a corporation; and
- had a complete set of accounts that enabled operating surpluses, savings and assets to be separately identified and measured (Hawksworth and Wilcox 1995: 52).

This, they argue, could be achieved through the ring-fencing of the current and capital accounts (and the current accounts are already ring-fenced in England and Wales).

Nevertheless, the authors argue that it would be better if a clean break were made and the stock were transferred to municipal housing companies. Stock transfer has proved to be more attractive to the government for public expenditure reasons. For some years, local authorities have been able under certain circumstances to transfer their entire stock of housing to a new landlord, usually a housing association. Not only does this have the advantage of taking future borrowing out of public expenditure, the public sector's stock of outstanding debt is reduced by the capital receipt received by the local authority from the successor landlord. The first call of the capital receipt is to repay outstanding housing debt.

Unsurprisingly, the strategy has been most straightforward in authorities where the value of the stock exceeds the outstanding debt, something much more common in rural authorities. But for most urban landlords, stock transfer is not viable without subsidy. Legislation passed in 1996 allowed local authorities in England and Wales to establish local housing companies and the government established an Estate Renewal Challenge Fund, allocated on a competitive basis, to facilitate the transfer of estates. It was notable that, amidst the large reductions in the social housing capital programme, the Government increased the size of this transfer fund for 1997–1998 and introduced other incentives to encourage stock transfer.

Because the starting point is different in the other member states, the British model does not seem to be transferable. The main exceptions are Ireland, where the relatively small social rented sector is mainly owned by local authorities, the city of Vienna, and the States of the former German Democratic Republic. Following unification, the social rented stock which had been directly in state ownership was transferred first to local authorities and then into municipal housing companies (McCrone and Stephens 1995: 64–65; Tomann 1996: 64).

In countries where housing debt is held by organisations outside the government sector, there is little gain in transferring this stock to other bodies. However, it is possible that the stock of outstanding public sector debt could be reduced where the public sector has ownership of social landlords through organisations such as municipal housing companies. If the returns on these companies were sufficient, the public sector could sell all or part of their share in the companies to private sector companies, in the same way in which governments have sold part or all of their share in nationalised industries and banks. In this way the government sector would receive a capital receipt which could be used to reduce other outstanding debt, although it may need to increase rents and hence expenditure on housing allowances to make the returns on the housing sufficiently attractive to new owners. This strategy has been widely discussed in Sweden, where, in 1996 the Housing Commission advocated the phasing out of interest subsidies on existing loans to municipal housing companies, and the introduction of one-off grants for new investment.

Ownership transfers illustrate a trade-off between current deficits and the stock of outstanding debt. The Dutch government, for example, has written off the subsidised loans of housing associations, but on the condition that there will be no future loan subsidies – an exercise known as 'grossing up' (McCrone and Stephens 1995: 90–91; Maclennan 1995: 14). This implies an increase in the government's stock of outstanding debt, but a reduced expenditure and therefore current deficit. The UK sold the Housing Corporation (public sector) loan book to a single buyer in 1997 (which, in turn, decided to securitise it), giving the government a one-off receipt, while Finland securitised some of the loans held by the Housing Fund in 1995 and 1996 (see Tulla 1996).

Conclusions

In the run up to 1998, when the founder members of the single currency will be selected, housing policies have been restructured in those countries which devoted the highest share of their national incomes towards subsidising housing in the earlier part of the decade. While the convergence criteria have not been the sole source of pressure on public expenditure, they have been an important factor and have added urgency to the restructuring of expenditure priorities. This chapter has also shown that monetary union implies an on-going requirement for members of the single currency to contain budget deficits and the EU has a number of sanctions, including fines, which it can use to exert fiscal discipline.

The budget-reducing strategies outlined in this chapter are therefore more likely to be the beginning, rather than the end, of retrenchment in the housing sector. Consequently, the search for efficient and innovative techniques in housing finance will be an important one.

Note

1 Whether the gains from devaluation are necessarily self-defeating is an empirical question. Nevertheless, some countries are now arguing for monetary union because of the competitive advantage gained by countries that devalued during the periods of currency turbulence in 1992 and 1993. This would indicate that, at least in the medium term, devaluations can have impacts on real economic variables.

References

Allsopp, C., Davies, G. and Vines, D. (1995) 'Regional Macroeconomic Policy, Fiscal Federalism and European Integration', *Oxford Review of Economic Policy* 11, 2: 126–144.

CEC (Commission of the European Communities) (1977) *Report of the Study Group on the Role of Public Finance in European Integration* (MacDougall Report). Brussels: Commission of the European Communities.

CEC (1990) 'One Market, One Money: An Evaluation of the Potential Benefits and Costs of Forming an Economic and Monetary Union', *European Economy*, 44 (Oct.) (whole issue).

EMI (European Monetary Institute) (1995) *Annual Report 1994.* Frankfurt am Main: EMI.

Hawksworth, J. and Wilcox, S. (1995) *Challenging the Conventions: Public Borrowing Rules and Housing Investment.* Coventry: Chartered Institute of Housing.

Hedman, E. (ed.) (1994) *Housing in Sweden in an International Perspective.* Karlskrona: Boverket.

Maclennan, D. (1995) 'Contrasting Fortunes: The 1990s Experience of Housing Associations in Western Europe', mimeo, University of Glasgow.

McCrone, G. and Stephens, M. (1995) *Housing Policy in Britain and Europe.* London: UCL Press.

MOPT (1992) *Informe para una nueva política de vivienda.* Madrid: MOPT.

Papa, O. (1992) *Housing Systems in Europe, Part II: A Comparative Study of Housing Finance.* Delft: Delft University Press.

TEU (Treaty on European Union) in B. Rudden, and D. Wyatt (eds) (1993) *Basic Community Laws*, 4th edn. Oxford: Clarendon Press.

Tomann, H. (1996) 'Germany', in P. Balchin (ed.) *Housing Policy in Europe.* London: Routledge.

Tulla, S. (1996) 'Securitisation and Finance for Social Housing: New Developments in Finland', *European Mortgage Review* (May): 21–31.

von Hagen, J. and Harden, I. (1994) 'The European Constitutional Framework for the States' Public Finances', paper prepared for the ESRC conference 'The Evolution of Rules for a Single European Market', University of Exeter, 8–11 September.

7

PLANNING, HOUSING AND THE EUROPEAN UNION

James Barlow

European Union involvement in housing provision has been described as a 'shadow' policy which is developing incrementally as a result of a wide range of legislation (Drake 1991). This includes legislation relating to the construction industry, employment, finance, the freedom of movement, social exclusion, and urban and regional planning. It is the last of these areas that is the focus of this chapter. European Union interest in planning has developed rapidly over the last decade and even though it is not directly concerned with planning for *housing development*, its regional development and environmental policy objectives potentially impinge on this activity. Whether the policies originating from the EU will lead to a gradual harmonisation of urban development planning systems, or whether the current diversity will be maintained, is unclear.

The chapter first examines the origins of EU interest in planning, since this helps to explain its focus on certain issues. Next, some recent trends in EU planning policy and the direction these may take in the near future are examined. Finally, we discuss the question of policy convergence and the possible housing policy implications of these trends for European approaches to urban development planning.

The scope of European Union involvement in planning

Current EU interest in national planning systems focuses largely on regional development and environmental issues, rather than on the implementation of specific urban projects. To understand the reasons for this emphasis, it is necessary to consider how the objectives of the EU have evolved.

As Goldsmith (1993) points out, arguably the most important objective of the EU – especially from the perspective of the European Commission (EC) – is the economic and political integration of its members. Much of the Commission's interest in urban and regional planning originally stemmed from its concern that regional disparities represented a threat to cohesion, especially after the admission of Ireland, Greece, Spain and Portugal. It is this concern which underpinned the European Regional Development Fund 1975, later reinforced by the Single European Act 1987 (Davies 1994; Davies and Gosling 1994).[1]

The primary purpose of the Single European Act was to pave the way for the single European market and a key feature was the reform of structural funds. Three principles underpin this reform; these establish the parameters for the EU's involvement in planning matters:

1 The concentration of financial assistance on priority objectives (focusing on regional development, combating long-term unemployment and integrating young people into the labour market).
2 The use of 'partnership' – taking a range of forms – involving the key parties in economic development and regeneration.
3 The establishment of consistent financial support strategies at local, regional and national level.

These principles clearly have implications for urban and regional planning. In particular, the EU is concerned that there is a clear planning framework, with firm development objectives, so that Structural Funds can be targeted to the best effect. This has resulted in calls to integrate better the different levels (regional and local; and economic, social, physical and environmental) of planning.

The SEA therefore gave the EU a firm mandate to intervene in regional and environmental matters. This mandate was reinforced in 1993 by Article 130s of the EC Treaty as amended by the Treaty on European Union (the 'Maastricht Treaty'), which stated that the Council of Ministers 'shall adopt . . . measures concerning town and country planning . . .'. Davies (1994) notes that this was the first mention of town and country planning in the Treaty.

The position of *housing* regeneration and development in this schema is not always evident, though. There have been proposals from the Committee of Regions for incorporating an 'urban competence' into the Treaty, to bring together the existing urban funding policies and mechanisms. So far, the Commission has resisted these proposals, although it has indicated that more could be done to raise the profile of urban issues. The URBAN initiative[2] is designed to provide financial assistance for innovative schemes to improve the quality of urban life. However, while the initiative is intended to target 'all elements which constitute present urban life', housing is a noticeable omission in the document. It seems unlikely that urban regeneration – including housing – will involve intervention at the European level unless it is perceived that there is a clear 'European interest' in relation to a core EU competence such as competitiveness.

While regional development – and more recently the question of social exclusion and the quality of urban life – have become important for European policy makers, it is the environment that has so far provided the Commission with more opportunities for actually implementing planning directives. Since the mid-1980s a series of Environmental Action Programmes have been put in place and the Directive on the Environmental Assessment of Projects was arguably the first piece of transnational planning legislation (Williams 1986, 1988). This paved the way for the current concern to promote sustainable development through more integrated action – the Fifth Environment Programme[3] is seeking greater integration of environmental assessment actions within the macro-planning process (Redman 1993). The programme was reviewed in 1996 and it was noted that progress towards implementation had been relatively slow because of inertia in 'attitudes'. Nevertheless, the Florence meeting of Heads of State in 1996 prioritised environmental protection and sustainable development. Although a draft directive on a highly comprehensive system for 'Strategic Environmental Assessment', put forward in 1991 and 1992, has not progressed to formal approval (Glasson 1995), clarification of what projects require environmental assessment is expected in 1997. Another environmental concern has been over the effects of strict zoning policies in the planning and development control systems of many EU states. An EC Green Paper on the urban environment[4] raised the question of the relationship between zoning and traffic generation and has suggested that wider consideration should be given to strategies emphasising mixed use and denser development. It therefore seems likely that the momentum within parts of the EU for greater intervention on environmental issues will remain powerful.

Emerging trends in European Union planning policy

There are three important limits to EU competence on planning matters, and Article 130s emphatically does not lay the foundations for a Europe-wide system of urban and regional planning. One reason is that the Article is subject to the unanimity rule, whereby all member states must agree before any action can be implemented. Another reason is that the principle of subsidiarity applies. This means that the Commission can only take action if individual states are unable to achieve the objectives of the Article themselves. This has been interpreted by a Director General for the Environment to mean that actions should be left to the local or national level, within a framework set by the Community. Indeed, Article 130t of the Treaty states that any measures introduced 'shall not prevent member states from maintaining or introducing more stringent protective measures'. Article 130s therefore represents an attempt to set minimum standards, rather than to ensure common action across all member states (Davies 1994). Finally, the reference to town and country planning in the Maastricht Treaty comes as an exception to the procedural rules governing the Environment Title. Thus planning is constricted within environmental policy and, according to Healey and Williams (1993), this raises the question of whether environmental and spatial planning can be regarded as a combined policy area or two separate areas which occasionally involve similar interests. There has, indeed, been tension between environment-led and regional-policy led planning movements in the Commission. Given the fragmented institutional structure of the Commission, this would seem to further impair attempts to define an EU planning strategy, even in a limited sense.

The Commission's principal statement on planning policy is the *Europe 2000+* report (EC 1994a). The report argues for stronger regional planning powers and improved integration of social, economic and environmental planning initiatives. *Europe 2000+* also summarises the spatial planning systems of the various member states. As Alden (1996) points out, this serves to highlight the importance of responsive and effective planning given the rapidly changing European and global context. A series of regional seminars was held in 1996 to discuss the report and a feedback report is being prepared for discussion at the June 1997 Inter-Governmental Conference (Nadin 1996).

Parallel to the work being carried out for *Europe 2000+* is the Committee on Spatial Development. This brings together ministers and civil servants responsible for planning and has been given the task of establishing a non-binding 'European Spatial Development Perspective' and a 'Trend Scenario' (Newman and Thornley 1996; Nadin 1996). The ESDP aims to promote a polycentric urban system, equal access of citizens to information and infrastructure and better management of Europe's 'cultural heritage'.

European Union planning initiatives and housing supply

What, then, are the implications of these planning initiatives for the supply of housing in EU countries? Given its current focus, it would seem that the immediate impact of EU planning-related policy will be on the *environmental* effects of housing development. We have already noted that there is a strong momentum to introduce more comprehensive environmental protection policies. And in the longer term, the intention to reduce commuting by promoting mixed-use development may have an impact on housing development in terms of its location.

European Commission directives to protect environmentally sensitive areas have already led to an overhaul of existing national planning systems (Ward 1995). For example, since 1994 planners have been required to assess the conservation implications of proposals affecting designated 'European sites' of environmental significance. In general, environmental policy is not applicable retrospectively[5] but in some instances (concerning European sites) directives are applicable to existing unimplemented planning permissions on those sites.

A secondary, less direct impact on housing supply will perhaps result from initiatives to improve the quality of life in run-down urban areas. In this case, the principal 'planning' effect relates to the need to integrate closely planning policies at all spatial and sectoral levels; unless this is achieved local authorities are liable to be disadvantaged in their bids for financial assistance for urban regeneration. In this way, local authorities may need to pay greater attention to the nature of housing development and rehabilitation, and its relationship to other sectoral issues such as transport and employment generation.

Another indirect impact of EU policy on housing supply may arise from the reform of the Common Agricultural Policy. This involves measures to reduce the intensification of farming and protect the countryside; as such it clearly has implications for land use planning and development in rural areas. In some circumstances the reforms may lead to an increase in land supply for housing in rural areas, as farmers diversify production. The lack of affordable housing in rural areas has been a major problem in much of the UK, for example, and CAP reform, coupled with new mechanisms for promoting cheaper rural housing (Barlow *et al.* 1994), have gone some way to meeting this need.

The effects of EU policy on planning systems will clearly vary from country to country, depending on their existing legal, political and institutional structures. In the UK the impact on housing development may be muted. One reason is that government policy is already moving towards the promotion of mixed-use development (Coupland 1996). Another reason is that a decreasing amount of new house-building involves large schemes on 'greenfield' land, which are most likely to be subject to environmental impact directives. European Union policies on environmental protection are therefore more likely to reinforce current trends than lead to a radical overhaul of the current system, although they may well spur a reluctant government into producing policy on, say, integrating transport and land use planning sooner than would have otherwise been the case.

The single European market, housing supply and land use planning

Healey and Williams (1993; cf. Davies 1994) point out that it is often assumed that the effect of the EU on national planning policies largely results from the activities of DG XI (environment) and DG XVI (regional policy). Also potentially significant are the implications of the macroeconomic objectives of the single european market (SEM).[6]

There has been some concern that planning systems represent a barrier to free trade, inasmuch as applicants for planning consent from one country generally need to seek expert advice in each member state. The mechanisms by which planning permission is granted can be somewhat opaque in many countries. There have therefore been calls for greater efficiency and responsiveness to market forces, as well as concern about the 'transparency' of property markets and 'certainty' of the planning process. Under Article 222 of the EU Treaty systems of property ownership are deemed to be matters for member states. Nonetheless, the EU seems to have signalled its intention to tackle the restrictions imposed by national systems of development control and planning on competition. One conclusion of the 1995 Inter-Governmental Conference was that:

the introduction of greater competition into many sectors in order to complete the internal market, should be compatible with the general economic tasks facing Europe, in particular balanced town and country planning . . .

(para. 1.6, Presidency Conclusions, Cannes Summit)

It is far from evident, however, that *housing* developers are particularly concerned about restrictions imposed on development outside their own country arising from obstructive practices in planning systems or property markets. Indeed, it is hard to envisage a situation in which competition between housebuilders is internationalising to any significant degree. Other than a few special cases (e.g. British developers building tourist-related housing in Spain and Portugal, contract house-building by German or Scandinavian firms in the former communist countries), housing development remains a resolutely national activity. Given the complex panoply of planning and building control regulations, subsidy systems and demand preferences, house-builders have so far shown little interest in Europeanisation strategies.[7] And even if a firm were to expand in another country it could simply buy the necessary expertise by taking over a local firm, thereby overcoming the problems of understanding different planning systems.

It is not surprising, therefore, that there have been no legal cases in which housing developers have sought to create greater transparency in the processes of obtaining planning consent. It is true that the validity of planning conditions designed to limit the occupancy of *commercial* and *industrial* buildings has been challenged under Articles 52 and 59 of the EU Treaty (which relate to freedom of establishment and freedom to provide services).[8] This does not, however, have an impact on occupancy conditions relating to *housing*. Healey and Williams's (1993) prediction that the EU may introduce regulatory codes which clearly specify the criteria against which development projects are judged therefore seems unlikely to be fulfilled.

Another area where the EU has the potential to affect housing supply is through public procurement legislation. This aims to increase transparency in procurement and improved market information in order to ensure that the conditions for competition are not distorted (Bovis 1993; Grover 1993). Public works and services include the planning and design of development projects, as well as their execution. The impact on housing development is, however, likely to be minimal since most schemes are too small to be subject to the legislation and do not involve the 'public' sector in any case. Some commentators have suggested that there may be a need to consider whether the Public Works Directive[9] is applicable to schemes where there is 'planning gain' relating to the provision of major infrastructure (Redman 1993). While large housing schemes in some countries can involve these arrangements, it is hard to conceive of a scheme that is so large that its infrastructure element would fall within the remit of the Public Works Directive.

Another significant factor behind the development of EU-wide housing policy would be greater intervention in housing finance systems (Pirounakis 1987; Ball

1990; McCrone and Stephens 1995). The likelihood of this occurring is unclear, but it is worth noting that when the proposals for liberalising the mortgage credit business among member states were being drafted, the EU explicitly wanted to distance itself from any hint that it might at that stage consider the formulation of a European housing policy (Pirounakis 1987).

Finally, DG III (the industry directorate) has turned its attention away from product conformity and setting standards (Priemus 1991) towards a more active policy in construction, in particular ways of using the industry as a generator of economic activity. This has largely taken the form of a vision of a series of mega-projects in infrastructure and transport ('trans-European networks') to boost Europe's overall industrial competitiveness and distribute the benefits of growth to less developed regions. Housing development is not a feature of this strategy but the agenda has indirect housing effects in terms of the second strand of the strategy, the improvement of competitiveness in the construction industry (EC 1994b). Of particular concern are the differences in specifications and design codes, and anomalies in contractual and professional practice. Some of these are being addressed by the Building Products Directive.[10] (See Chapter 6.)

To sum up, the emphasis of EU 'planning' policy has been on environmental control and improved integration of planning levels. The effects of this interest vary from country to country – for example, in the UK the impact of environmental legislation may become decreasingly important as the amount of housing development on greenfield sites declines. However, the comparative lack of integration in the British planning system may well conflict with EU objectives in this area. There has also been concern to open up planning systems and property markets under the single European market, but it is unlikely that housing developers will show any great interest in pushing for more rapid harmonisation. Of more importance for the building industry is likely to be legislation relating to construction and safety standards.

Convergence or harmonisation?

Is it possible to detect a 'convergence' in the broad planning policy trends experienced by different European countries; and to what extent does the legislation put forward by the EU play a part in bringing the planning policies of its members closer together?

Since the early 1980s there has been much discussion of policy convergence between states, prompted in the housing arena by an interest in the 'recommodification' of national housing systems (Barlow and Duncan 1994). The term 'convergence' conveys the idea that policies are becoming necessarily more alike, through some form of causal process. Convergence is felt by some (e.g. Kleinman 1996) to be more helpful than the concept of policy 'cycles', which emphasises the circulation and repetition of policies and fails to grasp notions of *dynamic* change.

It is by no means clear that there is any impetus towards a convergence of EU housing or local government systems (cf. Schmidt 1989). We can certainly observe some common themes. In housing policy these include (1) a weakened perception of the importance of housing as a national issue, as quantitative need has been met, (2) attempts by central government to reduce its share of direct housing costs, (3) attempts to devolve institutional responsibility for housing, (4) a shift to quasi-state service provision, a regulatory and supervisory role for central government, and an emphasis on 'customer' service delivery (Batley 1991). Furthermore, while the political strength of developers may have been adversely affected by central governments' views that the housing shortage is over, a concern to reduce the level of state expenditure may paradoxically have strengthened their position as the emphasis shifts towards public–private partnership in delivering housing and other forms of urban infrastructure. Under this schema, central and local government may take a more proactive approach to the planning and control of urban development, albeit one in which central government essentially takes a regulatory role. It is therefore possible that one outcome of the common themes experienced by European housing and local government systems will be a rediscovery of the importance of strategic planning for managing property markets. These common trends will, however, be mediated through the institutions, politics and histories of each country. There remain 'clusters' of policy approach, with housing provision grouped around distinctive finance, development and land use planning systems (Barlow and Duncan 1994; Harloe 1995; Kemeny 1992, 1995).

Will the EU's planning aspirations play a part in speeding up or reinforcing these trends, though? The outcome of the current deliberations for national land use planning systems is unclear. Some believe that the Commission's *Europe 2000+* proposals may largely deal with the strategic level, leaving the ESDP to tackle the problems of implementation at the local plan and development permit level (e.g. Morphet 1995). While the ESDP will be non-binding, it will nevertheless influence domestic policies on urban development and provide greater weight to land use and spatial considerations in decision making. The twin stimuli of *Europe 2000+* and the ESDP may therefore set the stage for a more closely integrated set of regional and sub-regional programmes. The two areas of special EU planning interest – environmental protection and the accurate targeting of financial assistance – may also provide support for a stronger role for strategic planning in some countries.

It is far from evident, however, that *harmonisation* of EU planning systems is a goal of Article 130s. And in any case, a major barrier to greater harmonisation is the local government framework within which planning is situated in each country (Healey and Williams 1993). In part, the opportunities to experiment with new forms of planning depend on the degree of local government autonomy and encouragement provided by national government. This varies significantly across the EU, with some local government structures much less conducive to the adoption of new approaches to planning. The British approach, for example, remains essentially reactive and the implementation of specific development

120

proposals is largely market driven. In the longer term the proposals in *Europe 2000+* offer the potential to remedy some of the UK planning system's current deficiencies, such as the poor integration of different forms of planning, short-term central government funding and limited democratic control.[11] Goldsmith (1993), however, feels there are major doubts as to whether the UK has a local government structure which is capable of responding to the 'new Europe' as envisaged by the EC.[12] He cites the distrust by local authorities of central government, the fact they have to deal with two central government departments (Environment, and Trade and Industry), the uncertain position of the new inte-grated regional offices and central government's interpretations of additionality and subsidiarity as factors which may all inhibit the UK from responding to proposals from the Commission.[13]

Conclusions

Can the planning aspirations of the EU be seen as a force bringing together diverse policy systems? It is *just* possible that the Commission will be concerned that local diversity will compromise its regional objectives and hinder cohesion, and this will strengthen tendencies towards the harmonisation of national plan-ning systems. Other factors which may prompt a gradual convergence include greater co-operation between European cities and regions, a general learning process, pressure to take into account environmental considerations and the need for integrated regional plans spanning international borders (Davies 1994). Fur-thermore, as Healey and Williams (1993) note, in the longer term EU policy papers have a significant role as campaigning documents, and rhetoric can turn into tangible policies and programmes.

However, it is more likely that an emphasis on devolution and diversity (the 'Europe of regions'), driven by a concern over subsidiarity, will prevail in mem-ber states. And further expansion of the EU may make detailed urban policy, focused on cohesion and social exclusion, even more difficult – disputes about who has or will benefit from the distribution of funds can only increase the prob-lems of finding policy majorities in the EU. Hence, a more probable scenario is for a convergence of *policy agendas*, but continued diversity in local planning systems.

We are not therefore dealing with a linear range of outcomes around some form of cross-national average. In that case theories suggesting convergence would still have a role to play, albeit a less deterministic one. Rather, we are dealing with distinct clusters of housing and planning 'regime'. These reflect the existing ways the EU is divided in terms of legal and administrative grouping (Newman and Thornley 1996), spatial divisions of labour, and capacities of national and local government institutions to make use of the Commission's programmes. These clusters will still predominate for the foreseeable future, albeit within a generalised EU framework of environmental and strategic plan-ning policies. *Housing* development is, however, less likely to be influenced by this framework than other types of development. In the absence of an express policy competence backed up by a history of action, housing supply in the EU will inevitably remain an outcome of other policy areas, although perhaps not a 'shadow policy' in its own right.

Acknowledgements

I would like to thank Peter Newman, Ken Armstrong, Ray Cocks and the editors for their helpful comments on an earlier draft of this chapter.

Notes

1 According to Hadjimichalis (1994), the EC fails to understand the complexities and the changing spatial division of labour in Europe and has therefore been preoccupied with overcoming the 'gap' between countries through an overly simplistic linear view of development.
2 COM (94) 61 Final/2: *Community Initiative Concerning Urban Areas (URBAN)*.
3 COM (95) 624.
4 COM (90) 218 Final.
5 The judgement in a recent case in Britain was that the Environment Assessment Directive did not extend to projects in relation to which the decision-making processes or development consents procedures had already started. See the case of *Twyford Parish Council* v. *Secretary of State for the Environment* [1992] 1 CMLR 276; (1992) 4 JEL 273.
6 See Wise and Gibb (1993) for a review of the general implications of the SEM.
7 There has been considerably more transnational activity in the contracting sector, but there are still complaints that construction markets for most projects remain largely local. Germany has recently been taken to the European Court for failing to ensure its construction markets are sufficiently open (Cooper 1994).
8 See Redman (1993) and Department of the Environment Circular 1/85, para. 74.
9 71/304/EEC as amended.
10 89/106/EEC as amended.
11 Although the EC-sanctioned delivery mechanism for specific schemes is likely to involve 'contracts' or 'partnerships' consisting of the main players, the degree of public participation in the decision-making process may well be relatively limited.
12 Arguably, this may not apply to Scotland because of its somewhat different local government and institutional structure.
13 Although he notes this problem is not unique to Britain – while research tends to draw attention to the pacemakers, most local authorities in EU countries are largely reactive to EC proposals and directives.

References

Alden, J. (1996) 'Regional Development Strategies in the EU: *Europe 2000+*', in J. Alden and P. Boland (eds) *Regional Development Strategies. A European Perspective*. London: Jessica Kingsley.

Ball, M. (1990) *Under One Roof. Retail Banking and the International Mortgage Finance Revolution*. Hemel Hempstead: Harvester.

Barlow, J. and Duncan, S. (1994) *Success and Failure in Housing Provision. European Systems Compared.* Oxford: Pergamon.

Barlow, J., Cocks, R. and Parker, M. (1994) *Planning for Affordable Housing.* London: HMSO.

Batley, R. (1991) 'Comparisons and Lessons', in R. Batley and G. Stoker (eds) *Local Government in Europe.* London: Macmillan.

Bovis, C. (1993) 'Construction and Planning Projects under the Framework of the EC Public Procurement Directives', *Journal of Planning and Environment Law* 816–822.

Cooper, P. (1994) 'Contractors Change Tack on Single Market Approach', *Construction News* 20 Oct.: 19.

Coupland, A. (ed.) (1996) *Reclaiming the City. Mixed Use Development.* London: E. and F.N. Spon.

Davies, H. (1994) 'Towards a European Planning System?', *Planning Practice and Research* 9, 1: 63–69.

Davies, H. and Gosling, J. (1994) *The Impact of the European Community on Land Use Planning in the United Kingdom.* London: Royal Town Planning Institute.

Drake, M. (1991) *Housing Associations and 1992.* London: National Federation of Housing Associations.

EC (1994a) *Europe 2000+. Co-operation for European Territorial Development.* Luxembourg: Office for Official Publications of the European Communities.

EC (1994b) *Strategies for the European Construction Sector.* Compiled for the European Commission by W.S. Atkins International Ltd, Brussels.

Glasson, J. (1995) 'Regional Planning and the Environment. Time for a SEA Change', *Urban Studies* 32, 4/5: 713–731.

Goldsmith, M. (1993) 'The Europeanisation of Local Government', *Urban Studies* 30, 4/5: 683–699.

Grover, R. (1993) 'Public Procurement Policies in the European Communities: Implications for Urban Regeneration', in J. Berry, S. McGreal, B. Deddis (eds) *Urban Regeneration. Property Investment and Development.* London: E. and F.N. Spon.

Hadjimichalis, C. (1994) 'The Fringes of Europe and EU Integration', *European Urban and Regional Studies,* 1, 1: 19–29.

Harloe, M. (1995) *The People's Home. Social Rented Housing in Europe and America.* Oxford: Blackwell.

Healey, P. and Williams, R. (1993) 'European Planning Systems: Diversity and Convergence', *Urban Studies* 30, 4/5: 701–720.

Kemeny, J. (1992) *Housing and Social Theory.* London: Routledge.

Kemeny, J. (1995) *From Public Housing to the Social Market. Rental Policy in Comparative Perspective.* London: Routledge.

Kleinman, M. (1996) *Housing, Welfare and the State in Europe. A Comparative Analysis of Britain, France and Germany.* Cheltenham: Edward Elgar.

McCrone, G. and Stephens, M. (1995) *Housing Policy in Britain and Europe.* London: UCL Press.

Morphet, J. (1995) 'Positive Planning, Courtesy of the Continent', *Town and Country Planning* 64, 5/6: 154–156.

Nadin, V. (1996) 'Spatial Planning in Europe. A Summary of Progress in 1996', mimeo. Bristol: University of the West of England.

Newman, P. and Thornley, A. (1996) *Urban Planning in Europe. International Competition, National Systems and Planning Projects.* London: Routledge.

Pirounakis, N. (1987) 'A European Market in Mortgage Credit', *Centre for Housing Research Discussion Paper* No. 15. Glasgow: University of Glasgow.

Priemus, H. (1991) 'Unification of the European Building Market: Possible Consequences for the Dutch Construction Industry', *Netherlands Journal of Housing and the Built Environment* 6, 1: 35–45.

Redman, M. (1993) 'European Community Planning Law', *Journal of Planning and Environment Law* 999–1011.

Schmidt, S. (1989) 'Convergence Theory, Labour Movements and Corporatism: The Case of Housing', *Scandinavian Housing and Planning Research* 6, 2: 82–102.

Ward, C. (1995) 'The Implications of New Directions', *Planning Week* 3, 15: 17.

Williams, R. (1986) 'The EC Environment Policy, Land-Use Planning and Pollution Control', *Policy and Politics* 14, 1: 93–106.

Williams, R. (1988) 'European Spatial Planning Strategies and Environment Policy', in G. Ashworth and P. Kivell (eds) *Land, Water and Sky. European Environment Planning.* Groningen: Geopers.

Wise, M. and Gibb, R. (1993) *Single Market to Social Europe. The European Community in the 1990s.* Harlow: Longman.

8

THE CONTROL AND PROMOTION OF QUALITY IN NEW HOUSING DESIGN

The context of European integration

Valerie Karn and Louise Nyström[1]

Housing design is, to a greater or lesser extent, the subject of regulation[2] by individual governments in Europe, involving a wide diversity of definitions of quality and of regulatory structures. In most countries there are building regulations which cover at least basic health and safety aspects of housing construction in all sectors, and increasingly go beyond this into energy and acoustic performance and even into areas such as accessibility for disabled people. In addition to the building regulations there are normally much more comprehensive types of housing quality 'norms', covering aspects such as internal space, layout and amenity. These 'norms' can be expressed in legislation, in mandatory building regulations, in official standards or as requirements of financing or insurance agencies. They sometimes cover all sectors, but more often apply only to housing in receipt of state loans or subsidies or within certain cost or size limits. In relation to the external environment and location, all countries have planning legislation which applies more or less to all sectors alike. In this chapter, we are mostly concerned with the first two types of regulation, but we make reference to the third as an important element in the whole picture.[3]

At any one time the nature of regulatory regimes is the product of the interaction between many international, national and sectoral forces, including existing and changing regulatory cultures and paradigms, technological advances and institutional structures (Dyson 1992a: 4; Francis 1993: 96). In the 1980s and 1990s, there have generally been substantial changes in regulatory regimes in Europe and America, and the regulation of housing design has not been unaffected. It is possible to identify a number of international influences on regulatory change in housing design in Western Europe, though with varying impact in different countries. These influences have been:

1   the transatlantic ideological swing towards deregulation as part of a wider
    project to 'roll back the state';
2   the deregulatory influence of the EU, related to:
    (a) attempts to harmonise rules to facilitate free movement of goods and
        services;
    (b) the changing style of technical standards as a result of attempts at
        harmonisation/mutual recognition;
    (c) the impact of fiscal policy on expenditure on housing, partly as a product of
        economic and monetary union and partly a product of national policies;
3   the growth of regulation related to energy, natural resources and the
    environment, stemming originally from international and national move-
    ments independent of EU initiatives but encouraged by specific EU
    directives;
4   the impact of international and national disability movements in increasing
    regulation on access for independent living.[4]

We will consider each of these influences in turn, before moving on to discuss
specific features of regulatory regimes for housing design in Europe.

The international (transatlantic) ideological shift towards deregulation

One of the major influences on the regulation of housing production has been
the wider ideological shift in views about the role of the state, and the increasingly influential critique of regulation as undermining economic efficiency,
innovation and competitiveness. This argument, which has been a recurring
theme of the Reagan, Bush, Thatcher and Major regimes (Eids and Fix 1984;
Goodman and Wrightson 1987) has been applied to regulation of housing
design as follows:

1   consumer preferences and choices in housing are best expressed by market
    demand;
2   overriding market demand in favour of predicted future value to society
    (whether through controls or financial inducements), is paternalistic,
    assumes shared values, ignores personal preferences, and risks wasted invest-
    ment in property people do not like;
3   enforced minimum standards push up costs to producers, current tenants
    and purchasers or, if subsidy is involved, to the state; much over-regulation
    results from the attempts of private interests to secure 'regulatory benefits';
    self-regulatory agencies are better informed and more likely to be trusted by
    the 'industry' than public agencies; relaxation of cost (rent) controls means
    that there is less need for quality control; (from design professionals) regula-
    tion distorts design decisions and restricts use of new technological solutions
    and innovative design ideas.

This critique, combined, as we shall see, with other deregulatory forces, has been most influential in changing the regulatory regimes in North European countries, such as Britain, Norway, Sweden and the Netherlands, where the old social democratic 'certainties' were coming under siege from market pressures and public disillusionment and where left-of-centre governments have been replaced by more conservative regimes or coalitions (Lundqvist 1992). In the adverse economic conditions of the late 1980s and early 1990s, such governments have had both ideological and economic imperatives to reduce state intervention.

However, this 'deregulatory' international environment has had a far less profound impact in some other countries, notably Germany, where the classic arguments in favour of social regulation have continued to receive widespread support. In relation to regulation of housing design these arguments have been:

1   *the public good,* such as public health and safety and environmental protection;
2   *the protection of investment, notably public investment* (this includes concerns about longer-term value-for-money, the length of useful life of the property and its suitability to meet consumer demand);
3   *protection of individual consumers' interests.*

The resistance of Germany to international deregulatory pressures stems from its strong 'regulatory culture' . This has three elements. First, and fundamentally, there is the over-riding politico-cultural emphasis on 'public-regarding obligations' which transcend market realities (Dyson 1992a: 10). This expresses itself in the effort to create conditions which seek to encourage individuals and groups, notably in the private sector, 'to pursue their self-interests in ways that are consistent with the public interest' (Dyson 1992a: 9).

Second, this sense of public obligation is translated into regulatory culture by German legalism which 'enshrines the primacy of legal rights and procedure' (Dyson 1992a: 10). The tendency is to codify social relationships into legal form, producing an elaborate and formalistic legal framework of regulation. Third, this is reinforced by a pronounced respect for the norms of objectivity and technical argument rather than adversarial contest.

So, despite strong pressures from the market, from the EU and from unification, it appears that in most sectors, so far, there is:

> no clear signal that the German regulatory tradition had lost its grip on policy. Still typical was a style of co-operative regulation, exhibited in a preference for sectoral self-regulation . . . and a tendency for change to be informally negotiated with the main organized interests . . .
> (Dyson 1992b: 259)

This political background broadly explains the continuing regulation of housing design in Germany. Given the size of the unified German market, the strength and expertise of its standards organisations and the political importance of the country within the EU, this is a pattern that clearly has a wider European significance.

Deregulation associated with the EU

Housing was not included within the areas of responsibility of the European Commission, either under the Treaty of Rome (1958), the Single European Act (1987) or the Maastricht Treaty (1992). Under the principle of subsidiarity, responsibility for administrative functions is lodged at the lowest possible governmental level; accordingly, housing is treated as a national rather than a European competence. The Commission does, however, have powers in related areas, namely the procurement of building contracts, the construction industry and construction materials[5] and the environment (Davis: 1992). There are also relevant aspects of the management of the European Union's (EU)[6] Regional and Social Funds (McCrone and Stephens 1995).

So housing design is not the subject of EU directives. But this does not mean that European integration is having no effect within this field. The regulation of housing quality is not insulated from the wider economic and regulatory influences operating within the EU.

The impact of harmonisation

The EU's aim to harmonise regulations and technical requirements, in order to promote a single internal market with free movement of goods and services across national boundaries, has had a 'knock-on' effect on the national treatment of technical requirements such as housing design, which are not currently required to be harmonised.

The European Community first tackled technical harmonisation in 1968. At that time the approach was to replace national standards with standards that were regulated by the EC. Although 270 directives, each full of technical detail, were adopted between 1969 and 1985 (Woolcock et al. 1991: 41), this method of proceeding proved highly problematic because member states repeatedly refused to harmonise technical standards, often because they were afraid that the result would be the lowest common denominator. Governments were keen to protect interests and to preserve national regulatory regimes. Negotiations to reach agreement through consensus were very time consuming and standards were often out of date before they were adopted. In the 1970s 'the whole process of technical harmonisation at Community level came almost to a standstill' (Woolcock et al. 1991: 41).

Given these difficulties, an alternative, more relaxed regulatory approach was adopted in preparation for implementation of the internal market at the end of 1992. Under this 'harmonisation' is replaced with 'mutual recognition and equivalence', which simply requires that the rules governing a service or industry must be the same for domestic industries as for industries from other member states. What these rules actually are is largely left to the discretion of the member states. However, the European Court of Justice has the power to declare national legislation a barrier to trade and therefore illegal. Thus if a rule makes it more difficult for industries from other member states to compete in the supply of goods or services, the national government has to be able to demonstrate that there is a genuine risk to health and safety or the environment and that the rule is not just being used to protect local industry.[7] Thus legislative harmonisation in technical standards is limited to health and safety and environmental consider-ations, since there has to be a definition at the level of the EU, or, failing agreement, at the European Court, of what constitutes a health or safety or environmental risk. In effect, the new approach involves, typically, harmonisation of minimum standards. This approach has been much more effective than the old one in removing non-tariff barriers to intra-EU trade and has in practice led to considerable deregulation in many areas of goods and services (Francis 1993: 34). As Bulmer says, this new emphasis:

> represents a transformation in the EC's regulatory instruments. This is enhanced by changes to the procedures of EC regulation, in particular the increased use of qualified majority voting. Taken together, the changes in the EC's regulatory goals and instruments can indeed be regarded as a 'paradigm change'.
>
> (Bulmer 1992: 63)

The major challenge for the EU is to achieve this freedom of movement but at the same time to 'avoid harmonisation that sinks to the lowest common denominator' (Francis 1993: 168).

Housing is not currently included within this regulatory framework. We specu-late later on the possible implications of its being included in the future. How-ever, it is possible that even though housing is not included, the existence of free movement of goods and services across frontiers is already having some indirect effect. For instance, it is claimed that one of the motives for deregulation of design requirements in Sweden was the desire to ensure that the Swedish house-building industry was not excessively geared to a highly regulated style of house-building, and therefore unable to compete elsewhere.

The 'new approach' to technical standards

The problems associated with the old style of directives also led to attempts to simplify technical requirements. A 'new approach' to technical standards has been adopted for EU directives, which is increasingly also adopted in national requirements, including those for areas outside the EU's competence, such as housing design. In this 'new approach' a distinction is drawn between 'essential requirements' formulated in terms of a 'general clause' in a directive which would provide the basis of legal enforcement, and the 'methods of satisfying the essential requirements', which are the technical specifications.

Compliance with the technical specification of an EU-recognised national or international standards body is the most convenient, but not the only way of meeting such 'essential requirements' (Woolcock et al. 1991: 45). At the national level a variety of types of organisations is involved in setting technical standards, some sectoral, some cross-sectoral, some semi-private/semi-governmental, e.g. BSI (UK), UNI (Italy), AFNOR (France), and some private, such as DIN (Germany), NNI (Netherlands) (Woolcock et al. 1991: 50). In practice the standards institutions regularly confer about the design of standards, both with each other and with the Commission. Where harmonised European technical standards are to be agreed, this is done within the European Standards Committee (CEN), which is a private sector organisation but which has adopted the same qualified majority voting system as the EU Council of Ministers, modified for EFTA membership.[8]

This 'new approach' to technical requirements is one that has been supported strongly by both the UK and Germany because it fits well with their philosophy of regulation. For instance, in Britain, the Building Regulations had already been amended in 1985 to include the concept of 'functional requirements'. The regulations also referred to 'guiding documents' which could be found in the EC's building-products directive. The 'new approach' also fitted with the British policy of deregulation (see below). Germany also supported the change, because the use of general legislative statements backed up with technical documents is one that is general in German regulatory systems:

> [T]here are strong elements of systemic congruence between the method of regulation being pursued by the EU and that which has been followed in Germany. The extent of congruence is greater than that between the EU and UK practice.
>
> (Bulmer 1992: 74)

As we just mentioned, though housing is not included in EU technical standards, there has been a widespread trend within member countries towards replacing the old detailed technical standards on housing design by this same approach, with legal requirements limited to general statements of 'essential or functional requirements', which refer on to technical volumes, providing guidance about the ways in which the essential requirement can be fulfilled. For instance, in Britain, the Housing Corporation abandoned the old 'Parker Morris' space standards in favour of a series of 'functional requirements', referring to the Building Research Establishment's *Housing Design Handbook* (BRE 1993) and subsequently to the National Housing Federation's *Guide to Standards and Quality* (NHF and JRF 1997). In Sweden, the old statutory dimensional standards have been replaced by functional requirements. In the Netherlands, the shift to using functional standards related to 'living areas' rather than rooms has particularly pleased the design professionals.

The approach has, however, attracted some criticisms. In particular, it needs to be recognised that the approach has a 'deregulatory bias'(Woolcock *et al.* 1991: 57; see also Bratton *et al.* 1996 on 'the race to the bottom').

> The stress laid on the formulation of essential requirements in the direct-
> ives is clearly an attempt to regulate only what is absolutely necessary at
> the European level and to apply the principle of subsidiarity.
>
> (Woolcock *et al.* 1991: 57)

This may or may not be what is desired but, in any event, it should not be a source of surprise after the event. The deregulatory bias is most pronounced if the functional standards are expressed in very broad terms. In these circumstances, it may be very difficult to obtain compliance. There was an example of this problem during the period of deregulation in Norway. The Housing Bank found that poor standard homes were receiving building permits because the functional requirements in the building regulations (e.g. 'suitable daylighting') were expressed in such general terms that they were difficult to enforce. On the other hand the scheme whereby supplementary loans were given for so-called 'life-span' homes was working well, because the quality requirements were much more precisely specified. Similarly, in England, the Housing Corporation's guidance was criticised as being too vague and 'allowing room for too much variance' (Stungo 1997: 10). Increased specification of functional requirements introduced in 1995, in response to criticisms about falling space standards, are said to have resulted in an increase of 12 per cent in the size of housing association property since the NHF's new guidance (Stungo 1997: 10).

Fiscal policy and the requirements of economic and monetary union

As Mark Stephens explains in Chapter 6, the Treaty on European Union has powerful implications for housing expenditure. Both the convergence criteria for monetary union and the broader aims of economic union require member states to observe 'fiscal discipline', that is, they need to control deficits. Taxation can play only a limited role in reducing deficits, so the overwhelming emphasis is bound to be on reducing public expenditure and government borrowing. For reasons explained by Stephens, housing has been and remains a major target for expenditure cuts.

There are, of course, other political and economic motives for cutting public expenditure and it is not always easy (or even possible) to disentangle them. Whether fiscal restraint is engendered by the EU or national policies, or a combination of the two, it may well, for a number of reasons, have an adverse influence on the quality of design of new housing.

First, there is likely to be pressure away from state funding and state provision of housing. In a number of member countries, control of housing quality is much more stringent for state-provided or state-subsidised housing, so the reduction of this sector means reduction in control. This, depending on the nature of control and the market situation, may more or less adversely affect housing quality, particularly for lower income groups.

Second, where housing continues to be state-provided or state-subsidised, the tendency will be for budgets to be cut and for there to be a downward squeeze on housing quality, through reducing capital costs via lower minimum standards and by cutting subsidies, whether 'bricks and mortar', tax relief or housing allowances.

Third, changes in the style of housing expenditure may have a crucial effect on housing quality. This particularly applies to the general trend in the 1980s in Europe away from bricks and mortar subsidies towards means-tested housing allowances, partly as a way of trying to reduce housing expenditure by focusing on the 'most needy'. One of the basic arguments for bricks and mortar subsidies has been that a country needs to subsidise the present generation in order to build to standards that will be appropriate in fifty or more years' time. This longer-term 'cost effectiveness' approach appears to be being discarded in most countries in a concern for short-term public expenditure savings. The argument is that the cost/rent of the property, and hence its quality, especially its space standards, have to be appropriate to a current tenant whose income is just above the qualifying level for housing allowance/benefit. This system tends to put downward pressure on housing quality.[9]

Deregulatory trends in housing design

The outcome of these various pressures has been a widespread move towards deregulation of housing design. This has been expressed both in the removal of regulations and in more subtle changes in the *character* of regulatory regimes.

The information approach to regulation of housing design

We have already discussed the 'new approach' to technical regulations. But there has been a broader change in regulatory regimes towards an 'information' rather than an 'interventionist' approach, that is for the state to concentrate on producing 'informed, autonomous consumer-citizens', protecting them 'simply by providing them with information about suspected dangers and letting them choose to accept or avoid the risk' (Francis 1993: 3).

The types of measures used are statutory or administrative requirements for minimum information (e.g. habitable area, orientation, energy rating, etc.), penalties for misinformation and inclusion of statutory warnings in advertisements. The information approach also lies behind the idea that consumers might be able to choose between products certified under different regulatory regimes. This type of 'indirect' approach to regulation ranges in type between a substantially regulated environment, where there is a requirement on housing producers to quote their performance against a wide range of official norms, right through to a much less regulated environment in which the government makes few if any information requirements, depending largely on competition between producers.

Britain belongs in the latter group. Though the government increasingly emphasises the sovereignty of the housing consumer, there is, ironically, no requirement on housing producers to provide minimum information about design, space and energy standards. The only requirement is that, when information is provided, it should be accurate (the Property Misdescriptions Act 1991). Even when there were mandatory minimum space standards in the public sector (1967–1982), there was no requirement on the private sector to produce information which measured houses against this standard. Nor was, or is there even a basic requirement to quote the habitable area of the property. As we will see later, there are however moves to develop a 'housing index', initially to measure housing quality in the social rented sectors but possibly, ultimately, to provide more information to home-buyers as well.

Sweden represents a clear example of a switch from an 'interventionist' regulation strategy to a thorough-going 'information' approach. Much of its system of mandatory minimum design standards has now been dismantled[10] and has largely been replaced by a system whereby properties are required to meet the functional standards in the Building Regulations. Examples of how these can be met are given in the 'Swedish Standard', but it is not compulsory to adopt these. Furthermore, it is the developer's obligation to enforce the building regulations as building quality inspection by the public sector has been discontinued. The expectation is that consumers will avoid buying or renting properties that fail to meet their needs. There are heated debates as to whether this will in fact be the case or whether quality will fall, and further debates as to whether this would matter.

The most common criticism of the 'information' approach to regulation is that it is less effective than interventionist regulation in protecting all groups in society, particularly the poorest and least educated. The predominantly middle-class status of members of consumers' organisations and readers of consumers' magazines demonstrates this point, as does the uneven effect of health warnings on cigarette advertisements on different age and class groupings.

There are, in addition, particular difficulties for consumers in exercising control over the quality of housing because of the complexity of the package of items involved. If a Swedish home meets some but not all of the Swedish standards, how will the potential buyer weigh up the significance of this? This assessment is all the harder because housing is an 'experience good', that is, you have to live in it for some time to know what is the effect of, say, poor storage capacity or poor sound insulation.

But the major problem is that lower income people, even if they are provided with an extensive description of the demerits of a property, have little option but to agree to rent it. And for lower income buyers, what the market has on offer has less to do with consumer preference than types of property that generate the most profit for the producers (Ball 1983: 143). As we shall see later, Norwegian experience in the 1980s demonstrated this problem in acute form.

Enforcement and compliance

A further variation on regulation relates to the system of enforcement. Public expenditure cuts tend to squeeze inspection and monitoring which can either undermine compliance or result in a greater emphasis on self-regulation or both. Instead of having their own systems of scrutiny and inspection, governments increasingly require self-regulation and self-certification, reinforced by spot checks and random audits. There is also in some countries a trend towards decentralised or privatised inspection. All these changes are often associated with a deregulatory trend, but are far from being fully fledged deregulation.

The degree of compliance is of course crucial. There is by no means a clear relationship between the directness of the system of enforcement and the degree of compliance. Indeed, it could be argued that those countries which have the best record of compliance find it possible to adopt much more self-regulation. The prime example is Germany, where there is 'an impressive level of compliance' (Woolcock *et al.* 1991: 51), but heavy reliance on self-regulation. One reason for this may be the strong emphasis placed on consultation in the design of regulations, which itself stems from the emphasis on regulation as a way of achieving 'public good'. As Bulmer and Dyson put it, 'German regulatory practice is based on an entrenched consensus with those who are regulated' (Bulmer 1992: 73) and, 'German attitudes towards regulation appear notably supportive. Regulatory action enjoys a high degree of legitimacy' (Dyson 1992a: 1). In contrast, in France, 'standards are less readily accepted by users and producers alike' (Woolcock *et al.* 1991: 51). It is clear that regulatory cultures are so different that two very similar regulatory systems could have very different outcomes in two countries, depending on the degree of compliance.

The opposite tendency – extension of regulation

Although deregulatory trends have been the most striking, there have also been extensions of regulation of housing quality in two respects: first, in relation to energy consumption, environment and natural resources and, second, in relation to access for disabled people. Neither of these stems from housing campaigns. They are by-products of two other quite distinct movements – the environmental movement and the 'independent living' movement.

Energy, natural resources and the environment

The European Community's environmental policy was initiated in 1972, in the wake of the United Nations Conference on the Human Environment in Stockholm in that year and in response to a surge of public concern about environmental deterioration. In the following twenty-five years, both the Commission itself and the member countries have developed more or less stringent environmental regulation. Those countries where environmental policies are particularly active have tried to use their influence within the EU to encourage international co-operation in fighting environmental deterioration.

Germany, in particular, promoted the 'principle of precaution' (*Vorsorgeprinzip*), stressing the need to anticipate rather than simply react to environmental problems. It was also forcefully argued that environmental protection should not be seen as being in conflict with economic development but that it, 'forms a necessary condition of such development. Thus high standards of environmental protection can form the basis for greater international economic competitiveness' (Weale 1992: 160–161). Both these principles clearly have relevance to the regulation of housing environments.

The EU has made considerable progress towards establishing uniform standards in some areas such as water quality and land-use but:

> if a member state wishes to adopt higher environmental standards than those enunciated by the Community, it may do so. In the Community's words, the stricter standards of a member state would win out over the less strict standards of the Community.
>
> (Francis 1993: 166)

This is a principle which would have relevance were housing ever to be included within the competence of the Commission.

A number of aspects of environmental regulation have a direct impact on housing design, notably those concerned with energy and water conservation in new housing, land-use densities, waste recycling, provision for public transport and the motor car and preservation of urban and rural landscapes. However, many of these are the relatively 'small beer' of national, state or even local-level regulation, rather than the subjects of EU action. The degree to which countries or individual jurisdictions within countries, have such regulations varies enormously. Most EU members now have some sort of legislation about energy and water conservation. In some countries energy requirements vary by region because of climatic differences. (There are also moves at the EU level to designate minimum insulation values for buildings in its various climatic zones.) Similarly water usage may be regulated according to climate and topography. In Sweden, water is metered, one reason being to keep down consumption. In 'dry areas' there is a legal requirement to give special consideration to water distribution and consumption. In England and Wales, where water supply has been privatised, the attitudes of different regional water companies to the installation of water meters in new housing vary, and government is suggesting regulating both in this respect and in respect of low capacity WCs and showers. In Sweden, provision for separating, storing and recycling different types of domestic waste is now being introduced in all housing and urban areas, something that in Britain has scarcely gone beyond the supermarket bottle bank. In the Netherlands and Scandinavia, care is taken with provision for cycle tracks and cycle storage. Planning controls to preserve the appearance of townscapes and countryside are universal, but financial incentives are less common; the Norwegian State Housing Bank, however, gives more favourable loans to schemes that are environmentally sensitive. Protection of potential housing residents from pollution, such as methane, dust and noise are also factors which have a much higher profile than they did twenty-five years ago, and, in Sweden particularly, there is widespread concern about internal pollution and the use of noxious building materials.

There is therefore a strong momentum behind environmental issues in housing developments, affecting amenity, cost-in-use and the appearance and health of the external setting. The increasing regulation of these aspects of housing contrasts, as we will see, with the widespread deregulation of internal space and design.

The impact of the Independent Living Movement

Over the last two decades, campaigns by and on behalf of disabled people have had an increasing impact on regulations about the design of housing. There has been growing rejection of the 'medical model' of disability and adoption of a 'social model' in which the 'problem' is identified not as the physical disability itself but the reactions of society to the disabled person. This has led to an emphasis on the rights of disabled people, and, in particular, their right to 'independent living',[11] and to 'normal' working and living environments rather than ghettoisation.

Accessible buildings, including accessible housing, have been key elements in the independent living campaign, along with accessible transport and equal opportunities in employment. There is a parallel between the types of arguments given in favour of investment in independent living and the economic arguments given for environmental protection (see above). In support of rights legislation for disabled people, proponents argue that it 'adds wealth by reducing their dependence on benefits and increasing the contribution they can make' (*Hansard* 1993: 1142, quoted in Davis 1996: 131). Possibly for this reason, on the whole much more attention has been paid to rights of access to employment than to housing; most EU member states have laws relating to the equal treatment of disabled people in employment.

The ageing of the population and the growing preference for community care rather than institutional care have reinforced the need to provide housing which facilitates 'independent living'. But the success of 'independent living movements' in relation to housing has been very variable. Sweden, Denmark and the Netherlands had early schemes of this type[12] and have continued this record with stringent requirements for wheelchair access in new or renovated housing, or at least a proportion of it in each development. The Swedish building regulations introduced handicap requirements in 1975, requiring all new dwellings to be designed with regard to use by people using a wheelchair. In contrast, despite the fact that Britain has contributed strongly to research on the design of accessible housing, construction has been almost invariably in grouped 'special needs' housing schemes (Barnes 1991: 149). There has been little commitment to ensuring that new housing as a whole is accessible or adaptable for 'life-time' use. As Barnes writes:

> There are no government directives to house-builders to build more accessible homes, nor have housing policies been proposed which encourage public and private landlords to convert existing stock.
>
> (Barnes 1991: 150)

However, over the last few years there has been much more pressure for change. Both the Housing Corporation and Scottish Homes, which regulate the housing association sectors in England and Scotland respectively, have recently introduced requirements for accessibility, and the National Housing Federation is seeking to make these requirements more stringent and applied to a larger range of dwelling-sizes. There was also an initiative to include accessibility requirements in the building regulations for new residential property of all tenures but it was blocked by opposition from the private house-builders. Since there are no special grant rates attached to accessible general needs housing, this failure to extend requirements to the private sector means that the cost of providing new accessible housing falls almost entirely on the rents of social housing tenants. It also means that disabled people have very limited choices of tenure and locations.

It is apparent then that the disability movement has had a marked effect on regulation of housing quality particularly in Scandinavia and the Netherlands, creating greater regulation at a time when the trend has been in the other direction. Requirements for wheelchair accessibility potentially have profound implications for the size and dimensions of rooms, kitchens, bathrooms and passages, the provision of lifts and for certain amenities, notably the provision of downstairs WCs in two-storey houses. While, on the one hand, general regulations about space standards are being relaxed or removed, accessibility requirements restore or even enhance these. They may, though, redistribute space from rooms to circulation areas in a way that would not suit all occupants. This is, for example, the case in Sweden.

Existing regulatory practice in Western Europe

Coverage

The aspects of housing quality that are the subject of government policy vary from country to country, as well as over time and with general political views. Generally, those aspects of housing quality which relate most clearly to wider public interest, particularly 'public health and environment', are the most likely to be controlled, whilst those that relate to protection of public investment and consumers, that is the interests of the actual occupiers, present or future, are less commonly controlled and are more variable in type.

Minimum health and safety standards in housing are regarded as indispensable. All EU countries have strict regulations on construction stability, sanitary and other environmental health precautions, dampness, ventilation, daylight, prohibition of toxic materials, thermal and acoustic insulation, fire precautions and the removal of waste. Minimum provision of bathroom, toilet and laundry facilities is also a rule. The fact that these topics are invariably the subject of regulation does not, however, mean that there is any uniformity of definition. For instance, fire precautions in apartments in the UK are very different from those in, say, Sweden and have a marked effect on the design of entrances. Similarly, stairs in the Netherlands are allowed to be much steeper than in the UK, and in France a WC compartment does not have to contain a wash basin as it does in the UK and Scandinavia.

But, as we saw earlier, ideas about 'the public health' have developed beyond the old sanitary and safety concerns. Concern for the environment and natural resources has introduced a major area of increased regulation, mostly concerned with energy and water supply and pollution but also protection of landscapes and other less quantifiable aspects of environmental protection.

Though there is considerable diversity of approach in the regulation of public health and environmental aspects of housing, once one moves beyond this to quality measures which relate more to the usefulness and comfort of the home, and the protection of investment and of consumers, the degree of diversity becomes much greater. The sorts of topics covered are the size, arrangement and functions of rooms, storage and circulation spaces, occupancy levels, external facilities, etc. These aspects of housing quality have been much more subject to the deregulation which has been a feature of European regulatory systems in the 1980s and 1990s. The exception, as we have seen, has been access for people with disabilities.

Diversity of national regulations

European housing demonstrates a mixture of the diversity stemming from vernacular housing and the much greater standardisation across frontiers produced by post-war industrial production. The latter was based on the common background of ideas about house design which originated in the experimental design work of the Bauhaus in Germany and then transferred to countries all over Europe. But even in products of relatively standardised appearance there are innumerable differences of detail, relating partly to the housing quality 'norms'[13] applied in different countries. It is far beyond the scope of this chapter to discuss the full range of diversity in these norms. However, we will illustrate the situation by a few examples.

Bedroom sizes

Minimum bedroom sizes were originally viewed in relation to health and the provision of enough air during the night but today regulations more reflect the functions of rooms. In some countries, minimum bedroom sizes are still quoted in terms of dimensions and space but, as we have discussed earlier, there is an increasing trend towards stating standards in terms of functional requirements. These functional requirements are typically backed up with reference to the minimum furniture that would be expected to be accommodated. Though functional requirements of bedrooms may sound very similar, the translation into furniture requirements and minimum dimensions varies widely from country to country and even between sectors in the same country. For instance, the National House-building Council (NHBC) which is the central organisation for insuring private sector house production in the UK, recommends that a single bedroom should have space for a bed, a bedside table, a chest and a wardrobe. The Housing Corporation, which regulates social rented housing in England, adds a desk to the NHBC's furniture list. The German DIN-norms require a child's bedroom to have a bed, a cupboard, one additional piece of furniture, a chair and a working table, plus circulation space to serve as play space.

Differences in furniture dimensions also affect the sizes of rooms. For instance, the NHBC's single bed is 0.75 m × 1.9 m; the Housing Corporation's is 0.9 m × 2.0 m; in the German DIN-norms, it is 1.0 m × 2.0 m and, in the Swedish Standard, it is 0.9 m × 2.1 m.

The combination of furniture requirements, circulation space requirements and furniture dimensions means that there is great variability in the minimum size of bedrooms. As we said, many countries no longer express their minimum standards in terms of areas, but it is instructive to refer to the old dimensional standards to demonstrate the scale of these differences. For instance, in Italy, the minimum size of a single bedroom is 9 square metres, compared with 7 square metres under the old Swedish requirements. At the other extreme, until 1994, the NHBC used to recommend that nothing below 4.5 square metres should be described as a 'bedroom'. The minimum size for a double bedroom varies from 9 square metres in France and Switzerland to 14 square metres in Italy and Germany.

Living rooms

The treatment of living rooms is even more variable. In Wales, housing associations are required by the funding agency (Tai Cymru) to provide a dining room or dining area in the kitchen or living room which will accommodate a dining table and chairs to seat the maximum design occupancy of the dwelling, a standard which 30 per cent of English housing association new production currently fails (Karn and Sheridan 1994: 56–57).

Since Parker Morris standards were ended, some individual English housing associations have derived their own norms for living rooms, based on roughly the same standard. The size of the living room has depended on the number of bedspaces in the house (about 11 square metres for one bedspace and 17 square metres for seven bedspaces) and whether or not dining takes place in the living room, which adds a further 2 to 3 square metres to the space proposed.

In France, living room sizes are also set in relation to the provision of other principal rooms. In Switzerland there is no minimum size but a minimum width of 3.5 m is set for the living room and at least one habitable room has to be 15 square metres or more. Germany requires 20 square metres if the dining space is included in the living room area and 18 square metres if it is placed elsewhere in the dwelling.

In Sweden, until the deregulation of 1994, the minimum living room size in family dwellings used to be 20 square metres, with a width of 3.6 m. The living room was then assumed to have space for a three-piece suite and a dining table, which was required in the kitchen as well. With the new building regulations, two furniture groups are no longer required for the living room area and it is only necessary to provide one space for dining, which may be situated in the kitchen, the living room or elsewhere. No minimum dimensions are given, but it is recommended that furniture groups described in the Swedish Standard are used to demonstrate the usefulness of the dwelling for different purposes. The aim is to define spaces which give as much freedom as possible to the resident, by providing room sizes and dimensions that allow for normal activities and furniture sizes.

A more thoroughgoing attempt to open up less rigid dwelling layouts has been made in the Netherlands Building Decree of 1991, with its 'free layout' principle. The only requirement is that there should be at least one 'staying room' of at least 3.3 m × 3.3 m in dwellings of less than 50 square metres and 3.6 m × 3.6 m in larger dwellings, to provide space large enough for a sofa, a coffee table and some easy chairs, or any other type of furniture arrangement the occupant may wish.

Cultural differences

Variety between countries is not just a question of regulations but of cultural preferences and traditions. For instance, it is considered acceptable in Sweden, but not in the UK, for people to have to cross the living room to get from their bedroom to the bathroom or WC. In Sweden it is felt that guests should not enter the bedroom or bedroom area, which is considered private domain. But they do need to use the WC, so, if there is only one WC, it has to be located close to the entrance or at least in the public parts of the dwelling. In other countries there are rather different definitions of what is considered public or private. In Britain the separation of public and private areas is produced by building largely two-storey houses with the bedrooms and bathroom upstairs; guests are expected to go upstairs to use the WC, but additional downstairs WCs are becoming common in larger homes. Even more commonly in the private sector, en suite bathrooms are provided for the main bedroom, even in very small homes, leaving other household members and guests with access to a separate bathroom and WC, albeit upstairs.

In Sweden, because of the climate, great attention is given to the storage of outdoor clothes and shoes near the front door. In Britain, not only is this aspect neglected but storage space as a whole is minimal. In the private sector, lack of storage space is compensated for by use of the garage for this purpose. It is also generally assumed in the private sector that there will be a spare bedroom used almost entirely for storage, hence the acceptance of very small dimensions for such rooms. In social rented housing there are no garages and few spare rooms, but no better storage provision.

In Switzerland, space standards are related to assumed occupancy (Lawrence 1996). A new residential building must have a minimum floor area of 20 square metres per person, including all internal spaces. In Britain, though social rented houses are defined in terms of their occupancy, there are now no clear minimum space standards applied in this way and there has always been strong resistance from the private sector against even describing dwellings in terms of occupancy.

In Britain and Sweden, to reduce dwelling size, it has become increasingly common for circulation space to be combined with living areas. In Switzerland, circulation space has to be deducted from the stated dimensions of any room which provides passage to another.

This enormous diversity of approach clearly has major implications in any discussion about the extension of harmonisation to housing design regulations.

Harmonisation of housing standards?

Under the principle of subsidiarity, housing is likely to remain outside the Commission's remit. However, it is instructive to speculate on what the possible implications would be of the regulatory approach of the EU being applied to housing design.

First, it would follow that any design requirements which constituted an obstacle to cross-border trade by house-building firms would have to be discontinued unless the requirements were defensible on grounds of health and safety. The first question is therefore whether the variety of requirements just described would be regarded by house-builders as constituting a real obstacle to trade across national borders and whether they would lodge a complaint. If no complaints were brought, the national systems could all continue unchallenged.

On the one hand, it is conceivable that a complaint about a barrier to trade might be brought. There is even a precedent in that English volume house-builders successfully exerted pressure for the repeal of minimum room sizes in the Scottish building regulations, in order to accommodate the very small 'third bedrooms' in their standard house-types. It is certainly true that the bedroom furnishing requirements of Germany, Sweden and Italy would make it impossible for English house-builders to market their standard dwelling-types. Even in their larger property-types there would be third and fourth bedrooms which would not meet the minimum standards of a 'single bedroom'. The low-priced standard dwelling-types, including the apartments, would fail much more radically. And standard Swedish apartments, though they would be very spacious compared with those in Britain, would fail on British fire-regulations about entrances passing the kitchen door and Dutch houses would fail on the steepness of their stairs.

On the other hand, one might much more convincingly argue that, given the differences in style, preferences and property type adopted in different countries, it is highly unlikely that an English building firm would expect to be able to market its standard products in, say, Germany. They would expect to redesign in order to appeal to the local market. In addition, design requirements could be a small obstacle as compared with other aspects such as access to land, complications of local planning controls, etc. More fundamentally, though, the structure of the house-building industry and the relatively local nature of their operations might mean there was little demand for cross-border trading and hence for harmonisation of regulations. This is highly likely, since, in other spheres with a far less parochial institutional structure than the house-building industry, firms have not shown eagerness to try to enter other countries' markets. Large builders that were keen to enter other markets might be more likely to try to purchase a local builder. They would then have no incentive to seek reduction in trade barriers to that market (see Chapter 5).

But assuming someone did bring a complaint of an unfair barrier, what chance of success would they have? Would the EU or the European Court conclude that, say British fire regulations, or German bedroom furniture requirements or Swedish wheelchair access constituted a barrier to cross-border trade? The argument would presumably be that outside builders incurred greater costs than local builders in meeting regulations. But such an argument might not in itself be enough to argue the case. Different builders could be differentially affected depending on whether or not they had standard house-types and, if they did, the characteristics of these. Mere differences in regulations between countries would probably not be sufficient.

But assuming that a country did successfully argue that their house-builders could not compete in another market, would the 'offending country' be able to argue a coherent case for their regulations on grounds of health, safety or the environment? The case certainly looks doubtful. The actual differentials in minimum room sizes look very hard to defend on these grounds, even though the functional requirements could be defended. And even the justification for different fire regulations looks shaky. Swedish builders could argue that the Swedish record of deaths by fire was no worse than Britain's and that, if Swedes were capable of escaping from an apartment past a burning kitchen, then the British should be able to do so. Similarly, the Dutch could probably show that no more Dutch than British people died by falling down stairs. And the French do not in all likelihood suffer greater ill health from the lack of a wash basin in their WC compartments.

Such debates would have the tendency to argue out all regulation or produce as an EU standard a lowest common denominator, of the sort that countries feared from the earlier attempts at harmonisation and which brought those negotiations to a halt. Clearly there would be severe political repercussions from reducing everything to the lowest common denominator of basic health and safety factors. It seems unlikely that Sweden and the Netherlands would have to drop their wheelchair accessibility because UK builders were unused to providing for it. It seems much more reasonable to suggest that the same principle would be applied as in environmental regulation, namely that countries were allowed to impose higher standards than those required by the EU or each other. But these higher standards would have to have a genuine rationale and not be just blatant protectionism. They would need to be defensible on the 'principle of precaution'. This would tend to produce some convergence of standards and definitions through consultation between standards institutions about the rationale for their regulations. Such international consultation already occurs to a considerable extent, for example in relation to energy efficiency and design for wheelchair accessibility.

However, it is also likely that the process would be to continue the moves towards simplification of regulations, the use of functional rather than dimensional standards and towards substantial deregulation of those aspects of housing design for which the rationale is most vague. It would tend to be argued that significant cultural preferences in design could be adequately protected by consumer behaviour.

Deregulation – four experiences

Given the possibility of more or less rapid and fundamental deregulation, we will close by looking to see what lessons may be learnt from the experiences of four countries which have deregulated their design requirements over the last fifteen years. These are Norway, England and Wales, the Netherlands and Sweden.

Deregulation and re-regulation in Norway

The process of deregulation in Norway (Christophersen 1994) dates back to the late 1960s when it was argued that the building code and regulations obstructed design innovation. In response some statutory requirements, particularly for daylight and storage space, were discontinued in the 1970s. However, this had little impact because 80 per cent of Norwegian housing was financed by subsidised loans from the Norwegian State Housing Bank, which had its own rigorous and well-specified requirements and careful scrutiny of submitted plans.

However, in the early 1980s deregulation became a more strongly stated government goal and in 1983 the Housing Bank was required to remove all its regulations except the maximum qualifying cost and size. It was argued at the time that there was no need for control of minimum housing quality because the Norwegian public was used to a very high standard of housing, and would exercise its purchasing power to maintain standards at a suitable level. In addition, deregulation was expected to introduce greater diversity of design.

In line with this approach, after 1983, all applications to the Housing Bank which were for homes within the Bank's maximum size and cost limits, and which had received a building permit, had to be approved for a loan. Extra loans could be provided for 'life span' homes, which still carried specific requirements, including wheelchair provisions. This system continued until 1991.

At first, established standards seemed to be retained but building and land costs, especially in inner city sites, started to escalate. The Bank began to be worried about the quality of housing being produced and undertook evaluation studies. Contrary to government expectations, these showed that housing quality had deteriorated during the 'regulation-free' period and that there was greater standardisation of house-types rather than greater diversity. Developers were maximising the number of units per site by building more storeys and smaller dwellings with lower ceiling heights and were reducing the number of access stairs and lifts by using long access corridors. Sun and daylight conditions became worse because flats were oriented in only one direction, and sometimes the windows were too small in relation to the area of the room. Rooms were deep and narrow and were difficult to utilise and furnish. Typically the living room had to be crossed to reach all the other rooms. The kitchens were narrow with poor working space and very often lacked dining areas. They were often placed so that they received only secondary daylight at the back of the living room. Storage space became much smaller. The outdoor environment deteriorated with smaller areas for play and recreation and insufficient sunlight. Despite all these problems, properties still met the very vague requirements of the building permits, so the Housing Bank had to fund them. The only houses that were guaranteed to have satisfactory space and layout standards were those subject to the precisely specified requirements for 'life-span' homes.

For these reasons the Housing Bank introduced a new loan system in 1992. The new requirements for a basic loan were to ensure that a home had a certain minimum standard and met the basic requirements of different types of house-holds. Quality demands were expressed as performance requirements and examples were given of acceptable solutions (which could be replaced by other solutions if it could be demonstrated that they met the performance require-ments). Supplementary loans were given for 'life-span' homes, well-planned out-door environments and good architectural design, environmental protection and innovative management.

Because the Housing Bank effectively re-regulated to avoid further waste of state investment in poor quality housing, it has not been possible to tell what would have happened over the longer term in an unregulated Norwegian market and, in particular, whether lack of demand would have led developers away from this style of construction. The builders did have some difficulty in selling their products in the private sector and were rescued by some of the property being purchased for social renting. The homeless, with no choice, were therefore sad-dled with the property, rather than the private developers paying the price of their design decisions. But it is not clear whether lack of consumer demand would have raised the standards. Studies showed that the private sector occu-pants were very dissatisfied with the dwellings, but this was after the event and had not prevented the initial purchase. Assessment of the situation is also compli-cated by the fact that there was a slump in the housing market in the late 1980s, so it is difficult to tell how much of the failure to sell these homes could be ascribed to their quality and how much to the general slump.

Overall though, the experience of deregulation in Norway was that it failed to meet the country's housing policy aims or to satisfy consumer preferences and it wasted public investment. The attempt to rescue developers from the con-sequences of their own design decisions left some of the most vulnerable people with a poor standard of housing. In addition, the builders were not forced to learn from their mistakes.

The experience raises a number of important questions. The first is clearly about the management of deregulation. The Norwegian experience showed the danger of sudden, thoroughgoing deregulation, under a misplaced assumption that developers, who had only ever experienced a regulated market, would natur-ally react by building the type of housing stock that people would want to buy or rent and that would represent a good investment of public money. But one could not confidently say that the Norwegian problems were solely caused by the developers' lack of experience of an unregulated market, because there is a wider question about the impact of boom and slump on the quality of production. During periods of boom, builders can capitalise on the fact that consumers will rush to buy, and, if they choose, can sell inferior property. This is where builders' interests may clash with governments'. Developers produce what is immediately marketable, whether or not it will stand the test of time. Once they have sold the property, they have little interest in whether the purchaser can sell it or let it. The only control is damaged reputation, if there is publicity about its failure to sell. However, from a government's point of view, the existence of housing that people do not want to buy or rent is a waste of national resources and may cause environmental and social problems if the properties are concentrated in large estates.

England – the owner-occupied sector

Comparison can be made with the private sector in England, which has not experienced regulation of design, except through the requirements of the building regulations. These are essentially geared towards health, safety and structural stability, but are quite widely defined and more demanding than those in Norway.

This is clearly a more settled market, in terms of builders' assessments of public demand and a study of production in 1991/92 showed that the majority of housing production in the private sector is well above the space standards for social rented housing. However, a minority of the very cheapest housing is built at standards very considerably below the minimum standard that used to be in force for social renting up to 1981 (see below) and it is this housing that raises questions about the desirability of such low quality additions to the housing stock.

There have been periods when the private sector in England has tried to produce very basic standard homes to keep costs down to those prevailing in the second-hand market. Though less dramatically than in Norway, developers have made investment mistakes, in terms of dropping standards to woo the first-time buyer market. In the early 1980s, so-called 'starter-homes' of minimal size (but with very generous mortgages) were built to attract first-time buyers in a difficult market. However, these received very bad publicity when they proved very hard to resell and mortgage and as a result most house-builders pulled out of this market, without government intervening to impose minimum standards. But, of course, those houses remain as part of the housing stock. Although this production was not financed by government loans as that in Norway was, the buyers' mortgages were subsidised by government through the Mortgage Interest Relief at Source (MIRAS) scheme and therefore represent government investment in housing.

Deregulation (and potential re-regulation) of the social rented sector in England

Perhaps the most significant example of the effects of deregulation is that of the social rented sector in England and Wales. From 1967 to 1981, design standards in subsidised rental housing were subject to the Parker Morris space standards, which were published in 1961. These standards represented what was being achieved in virtually all of the public sector in the early 1960s. But as part of the Thatcher government's policy of deregulation, Parker Morris standards ceased to be mandatory for local authority housing in England and Wales in 1981, though, in any case, local authority construction was virtually ended at this time. Until April 1982, Parker Morris standards remained a requirement of the Housing Corporation in its regulation of housing association production. However, since then the Housing Corporation has not stated floorspace requirements, relying instead on very generalised functional standards in its Scheme Development Standards. They were open to very broad interpretation. For some years, though, both the Housing Corporation and the housing associations themselves denied that there were any problems of falling standards, despite growing evidence to the contrary. They maintained that the associations themselves would ensure quality was maintained.

Two studies have demonstrated the fall in floorspace standards in housing association production since deregulation. One by the National Housing Feder-ation (then NFHA) in 1989/90 showed that 53 per cent of housing association property was being built more than 5 per cent below Parker Morris standards (Walentowizc 1991) and the other, by Karn and Sheridan, showed that this figure had risen to 68 per cent for production in 1991/92 (Karn and Sheridan 1994). Karn and Sheridan's detailed study for the Joseph Rowntree Foundation (JRF) also revealed that the decline in floorspace standards had been accompanied by other design changes which are closely associated. Increasingly, living and circu-lation areas were being combined; in 1989, 31 per cent of properties were reported as having no independent circulation space; by 1991/92 this had risen to 60 per cent. There were also extremely poor storage space standards; less than 7 per cent met the Parker Morris storage space standards.

But perhaps most important of all, housing association homes were being built in a form which allowed little scope for adaptation or enlargement at a later stage. Even where houses were not built as terraces (rows), plot sizes were too small to allow garages or additional rooms to be built in the future. Internally rooms were too small for the lack of storage space to be remedied later. These houses were of lower amenity value than the low-rise houses built by local authorities in the 1960s and 1970s and were being built in ever larger estates.

The cause of this big drop in quality appears to have been a combination of the 'value for money' competition for grant allocations, whereby housing associ-ations compete with each other to offer schemes with the least call on public funds, and the failure of the Housing Corporation to stipulate any adequate 'quality floor' around which this competition was to take place.

Ultimately both the Housing Corporation and the NHF concluded that some sort of quality floor was necessary. The Housing Corporation, the private inves-tors and the housing associations were all influenced by fears about the future value of the properties if they proved difficult to let. The result has been that the NHF, with JRF support, has produced a *Guide to Standards and Quality* with the intention that the Housing Corporation will adopt its basic recommenda-tions as minima in all bids for grants.

The *Guide to Standards and Quality* is also being used to help construct a Housing Quality Index, a simpler equivalent of the French Label Qualitel, covering a wide range of features of housing (55 by the latest report, Stungo 1997: 10). The immediate use of the index would be to assess social housing bids for their cost-effectiveness, although there is no suggestion at present of increased grant levels for better quality, as in the Norwegian system. In the longer term, it is also intended that the index should be used by the private sector to help potential buyers to assess the relative quality of homes.

Although the process has not been as sudden and dramatic as in Norway, the social rented sector in England is effectively being re-regulated. In neither country has this come about because of any change in political control. It has resulted from the public and private sector agencies involved with investing in social rented housing becoming anxious about the impact of declining quality on the future value of the property and the value for money being achieved from public expenditure.

Deregulation in the Netherlands

In the Netherlands up to 1991 there was a system of municipal by-laws, which were based on a common model by-law and expressed as minimum quality speci-fications (e.g. minimum room sizes). The new requirements under the 1991 Building Decree were expressed as performance requirements which are elaborated through functional descriptions, a limit value and reference to NEN (Nederlandse Norm) standards. As a safeguard on acceptable conditions, Dutch regulations also specify an absolute minimum. They are therefore still much more specified than the Housing Corporation's requirements in England.

One of the aims of changing the regulations has been to allow greater flexibility of layout and use of space, the so-called 'free layout'. This means that regulations are related to the whole 'living area' of the dwelling, within which the characteristic activities of a household can take place. By 'living area' is meant the total floor area that is intended for division into separate rooms, other than toilets and bathrooms, technical spaces and common circulation spaces. Thus, no distinction is made between living room, bedroom and kitchen. There are, however, some minimum requirements, namely that a living area of between 30 and 37 square metres may not be divided into more than two rooms. And, as we described earlier, minimum furnishable areas for living room activities are specified. There are also requirements for lifts and wheelchair accessibility.

Via the new Building Decree, the Netherlands government therefore confines itself to laying down general minimum regulations. Decisions about dwelling quality are increasingly taken at the local level. Local authorities can agree or enforce private supplementary building regulations, which may involve dwelling size, layout and type, sound and heat insulation, material and construction specifications, accessibility and urban design character. Also in a number of municipalities subsidy is given by the municipality for compliance with particular supplementary building and environmental requirements.[14]

At present the changes are too new to say if higher standards will be the effect.

Already, during the 1970s and 1980s, because of financial stringency there had been a big decline in the size, equipment and finish of new social rented housing in the Netherlands, despite the stricter regulatory regime in force then, and more recently the Netherlands government has withdrawn from subsidising new social rented housing. Commentators believe that this will mean that in future housing for low income people will need to be produced through the filtering down of older housing rather than construction of new housing. It is believed that social housing providers 'will be most reluctant to build very austere cheap dwellings in view of the future value of these dwellings and the competition from cheap dwellings in the existing stock' (van der Heijden and Visscher 1995: 83).

Deregulation in Sweden

In Sweden regulation of housing quality has applied to all sectors since 1975 when the norms which had previously applied to housing built with state loans were extended to all housing. (In any case state financing accounted for 90 per cent of homes.)

Criticism grew, however, in the 1980s about the detailed system of regulation and the constraining influence it had on design. Several demonstration projects were carried out by the building industry to show that good homes could be built without the straitjacket of regulations. The projects were evaluated and debated but with little consensus about their success. However, with the change of government in 1991, and its general deregulation policy, the National Board of Housing, Building and Planning removed several of the detailed items in the building regulations and others were reformulated as performance requirements rather than precise measurements.

One of the goals has been that it should be easier to build homes of differing standards and so increase variation in the market. The building regulations are now generally formulated statements about the functions that a home should accommodate, with reference to the Swedish Standards as a touchstone for acceptable quality. The 1994 Swedish Building Regulations recommend that a furnishing plan of the dwelling (using Swedish Standard furnishing dimensions) is drawn to make evaluation of the proposed project possible. But it is essentially now left to the market to ensure that the Swedish Standard is a requirement, as there is no compulsion to adopt it.

But even so, after deregulation, requirements in all sectors in Sweden are still very much more demanding than in British state subsidised housing. The building regulations still require dwellings to be designed with regard to use by people using a wheelchair, which determines the size and dimensions of rooms, kitchens, bathrooms and passages. Furthermore, lifts are required in three-storey residential buildings. So despite 'deregulation', the Swedish building regulations are still amongst the most far-reaching in Europe.

Little is yet known about the impact of the new system in Sweden, although there is a suggestion that dwelling sizes are being reduced and certain features, such as balconies, are being cut out. As in the Netherlands, there is far less government interest in housing production than previously, the focus now being upon costs, finance, the general quality of neighbourhoods and combating social segregation. This lack of interest in production and the priority that is being given to cutting public expenditure raise questions about the degree of monitoring that there will be of outcomes and the attention that will be paid to the quality of inspection and enforcement. This is important because quite new systems of inspection have been brought in. Essentially there is a system of self-declaration by producers, for which they have to obtain inspection certificates from registered private agencies. The Swedish consumers, both owners and tenants, also have to become aware that they are the system of enforcement of the Swedish Standard.

Deregulation – diversity of experiences

It is apparent that there is as much variety in styles of deregulation as there is in regulation. Compared with the types of much more thoroughgoing deregulation that happened in Norway and England, Sweden and the Netherlands have not so much undergone deregulation as a profound change in 'regulatory style', with more emphasis on self-regulation, decentralised enforcement and consumer information and less on direct government controls and inspection. This is consistent with the political changes that have taken place in those countries. As Francis has commented, in the broader context of regulatory trends, 'While confidence in the institutions of the state to resolve critical issues may be at risk, interest in new regulatory solutions has never been greater' (Francis 1993: 260).

The style of deregulation in Sweden and the Netherlands has also been partly a reaction against the highly industrialised building of monotonous estates with uniform and standardised dwellings that typified so much post-war construction. It is hoped that relaxation of controls will introduce greater variety of design and more innovation. However, the experience of Norway and England makes it at least questionable whether the new systems in Sweden and the Netherlands will deliver this. Private builders are likely to build what is convenient for them and marketable. Deregulation in Norway and England produced greater uniformity of production rather than greater variety. Mass production of uniform house-types is now more of a feature of the private than of the public sectors.

The need for improved information and protection for house-buyers, tenants and social landlords is a theme in all four countries. In no country is there yet the right to sufficient information. Moreover, if the consumer is increasingly to have to exercise the main control on quality, there needs to be a stronger style of consumer protection and education, going beyond 'information' into 'health warnings'.

Conclusions

In this chapter we have shown that European governments are being subjected to a series of political and economic pressures which are much wider than housing policy but which have a profound impact upon the regulation of housing quality and more broadly upon approaches to providing decent housing for people of all incomes.

In reacting to these pressures, it is important that housing policy goals are not forgotten. In the context of this chapter, it is vital that the system of regulation adopted is appropriate to housing policy goals, not just consistent with a deregulatory or regulatory ideology. But defending policy goals against powerful external pressures is almost impossible if they are not well articulated and many governments fail to make their housing quality objectives clear or relate policies towards new housing to the conditions in existing housing. For instance, is new production supposed to improve housing quality, merely add to the volume of existing housing or add specific types? How long is housing currently built supposed to last and what implications does this have for design? How is decent quality housing to be delivered to even the poorest households?

Britain in particular seems to have been short-sighted in this respect for a decade and a half, which has made it vulnerable to erosion of housing quality. But the governments of other countries which have in the past had longer-term strategies, such as Sweden and the Netherlands, are moving away from government intervention in housing quality. What the impact of this is likely to be may take decades to appear, but our four case studies give some hints.

Most crucially, it appears that the contradictory pressures on government are tending to have the effect of de-emphasising the protection of housing quality for the poorest. While there is, correctly, an increasing emphasis on the long-term cost and non-financial effects of failing to protect environmental resources and external residential environments, the trade-off appears to be that protecting minimum internal design standards is now widely regarded as too costly. Thus, in the four countries discussed, there has been a tendency to deregulate internal space and layout (apart from access for disabled people) whilst increasing regulation of the external environment. It is not an exaggeration to conclude that those features which are the subject of increased regulation are those which are more likely to affect the more affluent as well as, or even more than, the poor, while basic features which affect only the lower income groups, such as space and storage, and balconies in apartments, have been exposed to market forces, reinforced by cuts in subsidies.

The claim that governments believe the market will provide adequately is not entirely convincing. There is for instance a contradiction between governments' recognition that the market is unlikely, voluntarily, to produce wheelchair accessible housing, and the belief that it will produce housing adequately designed for other relatively powerless consumers, such as low income social housing tenants and the homeless. Similarly, while there is a touching faith in a deregulated

151

system to produce variety and sensitivity in internal design, even the most 'deregulated' countries show little tendency to trust the private or public sector house-builders to perform satisfactorily, without controls or at least financial incentives, in relation to certain key features of external housing environments.

In this rapidly changing regulatory environment, European countries can benefit from the sharing of experience and ideas about the promotion of housing quality. There is already a degree of convergence in the policies and practices being formulated and this convergence is likely to grow as countries scrutinise and rationalise their regulatory arrangements. Some ideas need to be further developed, notably offering financial incentives for better quality, instead of just imposing sanctions on bad quality, and the provision of improved consumer rights to information.

Notes

1 The research is based on the work of the ENHR Housing Quality Working Group. The authors wish to thank all the members of the group for their contributions and Scottish Homes and the Department of the Environment for their financial support.

2 Regulation is commonly defined as 'state intervention in private spheres of activity to realize public purposes'(Francis 1993: 5). In this discussion of its application to housing, we include, however, where appropriate (for instance in Britain), the regulation by the state of its own activities at different levels and of government agencies of various types involved in the production of housing. This does not apply in Scandinavia where all housing production is performed by independent bodies or agencies and none by central or local government.

3 Within 'housing design' we do not include construction quality and materials, except those aspects that relate to regulation of energy and acoustic performance.

4 Technological, economic and social changes have also played their part but are not the subject of this chapter.

5 For example the Construction Products Directive (89/106 EEC) which is aimed at removing barriers to trade in construction products.

6 The term 'EU' is used for activities after 1993, the date at which the European Community (EC) gave way to the European Union and, for simplicity, when general remarks are being made about European integration. The exception is when 'EC' appears in a quotation from another author.

7 In the environmental field this requirement is problematic in that by definition there is likely to be a lapse of time before environmental damage can be proved (Francis 1993: 148).

8 The 'big four' standards institutions (DIN, AFNOR, UNI and BSI) have forty votes out of the EU members' seventy-six, while EFTA members have twenty votes.

9 Germany has moved against this trend to vary direct subsidies for new rented housing according to the general expected income levels of the future tenants. This may have the effect of simultaneously protecting quality and reducing the risk of creating areas of 'unemployment trap' (Maclennan 1995, quoted in Stephens, this volume).

10 This has been partly counteracted by the firm and detailed disability requirements.

11 The Independent Living Movement started in the USA in 1969 and was part of

the wider movement in the early to mid-1970s for disabled people to control their own lives and campaigns. There is now a European Network for Independent Living. It is closely linked with the wider movement for rights for disabled people. In 1982 the United Nations initiated the Decade of Disabled Persons (1983–92), emphasising the responsibility of governments to protect and promote the rights of disabled people.

12 Namely the Swedish Fokus schemes, Collectivhaus in Denmark, Det Dorp in the Netherlands and Centres for Independent Living (CILs) in the USA. At the same time that these were being developed, Britain was building 'young chronic sick units' (Davis 1993: 288).

13 By housing quality 'norms', for the purposes of this chapter, we mean any kind of official or semi-official statement about housing quality, regardless of the legal status of this statement.

14 These arrangements contrast with the much weaker planning powers of British local authorities, which have no legal powers to enforce internal design requirements, although they attempt to do so as a condition of the release of their own land to private builders or housing associations. For instance, Birmingham City Council has its own standard house plans, which partly reflect the needs of the Asian households living in the area.

References

Ball, M. (1983) *Housing Policy and Economic Power: The Political Economy of Owner Occupation*. London: Methuen.

Barnes, C. (1991) *Disabled People in Britain and Discrimination: A Case for Anti-Discrimination Legislation*. London: Hurst and Co.

Bratton, W.W., McCahery, J., Picciotto, S. and Scott, C. (eds) (1996) *International Regulatory Competition and Co-ordination*. Oxford: Clarendon Press.

BRE (Building Research Establishment) (1993) *Housing Design Handbook*. Watford: BRE.

Bulmer, S. (1992) 'Completing the European Community's Internal Market: The Regulatory Implications for the Federal Republic of Germany', in K. Dyson (ed.) (1992) *The Politics of German Regulation*. Aldershot: Dartmouth, pp. 53–78.

Christophersen, J. (1994) 'Alleviating Adverse Effects of Deregulation', ENHR Conference Paper, Glasgow: University of Glasgow.

Davis, I. (1992) *The Implications for Housing of Current and Proposed European Community Directives: The Technical Issues*. Amersham: NHBC.

Davis, K. (1993) 'On the Movement', in J. Swain (ed.) *Disabling Barriers – Enabling Environments*. London: Sage.

Davis, K. (1996) 'Disability and Legislation: Rights and Equality', in G. Hales (ed.) (1996) *Beyond Disability: Towards an Enabling Society*. London: Sage, pp. 124–133.

Dyson, K. (1992a) 'Theories of Regulation and the Case of Germany: A Model of Regulatory Change', in K. Dyson (ed.) *The Politics of German Regulation*. Aldershot: Dartmouth, pp. 1–28.

Dyson, K. (1992b) 'Regulatory Culture and Regulatory Change: Some Conclusions', in K. Dyson (ed.) *The Politics of German Regulation*. Aldershot: Dartmouth, pp. 256–271.

Eids, G.C. and Fix, M. (1984) *Relief or Reform: Reagan's Regulatory Dilemma.* Washington, DC: The Urban Institute Press.

Francis, J. (1993) *The Politics of Regulation: A Comparative Perspective.* Oxford: Blackwell.

Goodman, M.R. and Wrightson, M.T. (1987) *Managing Regulatory Reform: The Reagan Strategy and its Impact.* New York: Praeger.

Hansard (1993) Parliamentary Debates, 26 February. London: HMSO.

Karn, V. and Sheridan, L. (1994) *New Homes in the 1990s: A Study of Design, Space and Amenities in Housing Association and Private Sector Housing.* Manchester and York: The University of Manchester and Joseph Rowntree Foundation.

Lawrence, R. (1996) 'Housing in Switzerland: The Definition, Promotion and Control of Quality', ENHR Housing Quality Working Paper, Geneva: University of Geneva.

Lundqvist, L. (1992) *Dislodging the Welfare State? Housing and Privatization in Four European Nations.* Housing and Urban Policy Studies, 3. Delft: Delft University Press.

Maclennan, D. (1995) 'Contrasting Fortunes: The 1990s Experience of Housing Associations in Western Europe', Mimeo. Glasgow: University of Glasgow.

McCrone, G. and Stephens, M. (1995) *Housing Policy in Britain and Europe.* London: UCL Press.

National Housing Federation and the Joseph Rowntree Foundation (1997) *Guide to Standards and Quality.* London and York: NHF and JRF.

Stephens, M. (1998) 'Fiscal Restraint and Housing Policies under Economic and Monetary Union', in M. Kleinman, W. Matznetter and M. Stephens (eds) *European Integration and Housing Policy.* London: Routledge, pp. 97–112.

Stungo, N. (1997) 'Space: The Final Front Room', *RIBA Journal,* March, pp. 10–11, 13.

van der Heijden, H. and Visscher, H. (1995) *Housing Quality in the Netherlands.* Delft: OTB.

Walentowizc, P. (1991) *Housing Standards: A Survey of Space and Design Standards in New Housing Association Projects,* Research Report. London: National Federation of Housing Associations.

Weale, A. (1992) *The New Politics of Pollution.* Manchester: Manchester University Press.

Woolcock, S., Hodges, M. and Schreiber, K. (1991) *Britain, Germany and 1992: The Limits of Deregulation.* New York: Council of Foreign Relations Press.

Part III

OUTCOMES

9

HOUSING, TENURE AND INTERNATIONAL COMPARISONS OF INCOME DISTRIBUTION

John Hills[1]

Countries vary widely in their housing systems and in the ways in which they subsidise or give tax concessions to their housing. These differences affect the distribution of living standards within each country, as the benefits in kind which people derive from housing ('housing income') are unlikely to be distributed in proportion to other forms of income. They also affect comparisons of income distribution between countries. Because most comparisons are made between *cash* incomes, excluding some or all forms of housing income, problems will arise if, for instance, those with high incomes are more likely to be owner-occupiers (with significant imputed rents contributing to their standard of living) than those with low incomes, or conversely if those with low incomes are more likely to occupy subsidised social housing, where gross rents are below those which would be charged in the market.

There are substantial variations within the European Union in tenure patterns and the extent and form of housing subsidies. In some countries, with extensive private rented sectors unaffected by rent control, cash incomes would require little adjustment for housing income to give a picture of overall living standards. In others, with large owner-occupied sectors or with large social housing sectors renting accommodation to low income groups at below-market rents, housing income may be substantial in relation to cash incomes. Comparisons of aspects of income distribution, such as the proportion of the population with incomes below half the national average, may thus be seriously distorted if they do not allow for housing income consistently. This chapter is based on work carried out for Eurostat, reflecting concerns that such distortions might affect such comparisons between EU members, particularly given the extensive use made of comparisons of poverty rates based on measures excluding housing income.[2]

This chapter examines the theoretical issues involved in measuring and allowing for housing income, the potential effects of variations in tenure patterns within twelve EU countries, and variations in tax and subsidy arrangements for housing between the countries. The chapter then presents some illustrative estimates of the distributional effects of housing income on 'poverty' measures using microdata for two member states – France (1984–1985) and the UK (1989).

Allowing for housing income: principles

An ideal measure: comprehensive income

In order to understand the problems which differences in housing systems can cause for international comparisons of income distribution, it is helpful to start by considering how housing would be reflected in an economist's ideal measure of income, 'real comprehensive income' (Pechman 1977; Meade Committee 1978). There are three features of this measure which are relevant here:

1   It includes income in kind, as well as in cash. In the context of housing, an owner-occupier derives value from being able to live in a house without paying rent, and this value should be included in income in some way, if the living standards of owners and tenants are to be compared fairly. Similarly, a tenant who pays a below-market rent for some reason (such as subsidies to social landlords, rent control in the private sector, or accommodation provided free with work) derives a 'housing income' from the difference between actual rent and market rent.[3]

2   In principle, it also includes capital gains:

> According to this definition, income is the accretion of the power to consume. It consists of a person's actual consumption plus or minus any increase or decrease in the value of his power to consume in the future as measured by his net worth.
>
> (Goode 1980)[4]

Capital gains contribute as much to raising net worth and future consumption possibilities as flows of income like interest and dividends. Note, however, that the concern is with *real* capital gains, after allowing for the effects of inflation on asset values. 'Paper gains', simply reflecting the effects of general inflation, do not contribute to future consumption power. Conversely, real *losses* should be allowed for if inflation erodes the value of financial assets.

3   Interest *payments* should be deducted when calculating income. Someone who borrows ECU 1,000 at an interest rate of 7.5 per cent and invests it in assets yielding 10 per cent has an annual income of ECU 25, not ECU 100. Again, it is *real* interest payments which should be deducted, not the full nominal payments if real debt burdens are being eroded by inflation.

For owner-occupiers, comprehensive income would include the *net* imputed rent on their property, *less* real net interest payments on borrowing, plus real capital gains on the property (less any resulting capital gains tax). Net imputed rent is an estimate of the market rent the property could command, deducting expenses like repair, maintenance and depreciation.[5] Net interest payments would reflect the effects of any tax reliefs or subsidies benefiting owners with mortgages. Income would also include any income-related housing allowances received by owners.

For a tenant paying a full market rent without any assistance from housing allowances or tax reliefs there would be no housing income. If the tenant received a housing allowance, this would, however, contribute to income (and might or might not be included in conventional cash income measures). In addition, if tenants pay below-market rents, the difference between actual and market rent would also be part of comprehensive income. Note that it is the difference between actual rent and *gross* market rent which has to be estimated, since tenants do not normally have to bear the cost of repairs, etc. (although this may vary depending on the form of lease).

Practical applications: 'With Housing Income' measures

The advantage of the comprehensive income definition is that it allows a fair comparison between households in different circumstances and between countries where institutional arrangements vary. It is, however, an ideal, and available data may not allow its calculation in full. The two biggest practical difficulties relate to capital gains and mortgages.

The empirical exercise reported below was designed to illustrate the sensitivity of international comparisons of income distribution to differences in housing systems. Conventional measures of income distribution are usually based on reported *cash* incomes, rather than real comprehensive income. Such measures suffer from a number of internal inconsistencies. When allowing for housing incomes in kind a compromise has to be made in terms of which inconsistencies to tolerate, and which to avoid.

Capital gains

Income distribution statistics based on household budget survey data usually exclude any element for capital gains or losses on financial assets. It would be anomalous to include capital gains on housing when they were not allowed for on other assets. The 'With Housing Income' (WHI) measure developed below therefore excludes any allowance for capital gains.

Mortgages

Establishing a consistent treatment of mortgages is difficult, particularly if capital gains and losses are being ignored. In principle, four approaches could be taken, none of them ideal:

1   Ignore mortgages altogether, and credit all owners with the full net imputed rent. This is consistent with the conventional treatment of incomes from other assets, where borrowing costs are usually not allowed for. However, this creates anomalies between tenants paying full market rents and owners with 100 per cent mortgages. In the absence of inflation, anticipated capital gains or special concessions for mortgage borrowing, the outgoings of the owners would equal those of the tenants, and they would have no economic advantage. Crediting them with the full net imputed rent would therefore exaggerate the position of mortgagors relative to tenants.

2   Credit mortgagors with the net imputed rent, but deduct their nominal mortgage payments. In cash flow terms this gives a reasonable representation of the position of mortgagors, whose income available for other forms of immediate consumption is limited. However, it ignores the advantage to borrowers of inflation eroding the real value of their liabilities. The higher the rate of inflation, the 'poorer' the mortgagors will be on this measure, even though their long-run economic position is unaffected.[6]

3   Credit mortgagors with the net imputed rent, but deduct their *real* net mortgage payments (after allowing for inflation, subsidies and tax reliefs). This produces adjustments which are robust to differences in inflation rates. Implicitly it attributes all of the real mortgage payments as a 'cost' in generating the net imputed rent, but not as part of the cost of generating the expected capital gain (which is being ignored in line with gains on other assets).

4   Credit mortgagors with a net imputed rent calculated on their *equity share* (i.e. total value minus outstanding mortgage), together with an adjustment for any subsidies or tax reliefs. Implicitly, this divides real mortgage payments as a cost attributable to generating the net imputed rent and the expected capital gain, in proportion. In the absence of expected capital gains, and if net imputed rents are the same percentage of property values as the real rate of interest paid by mortgage borrowers, this approach gives the same answer as 3.

Whichever of these approaches is taken, there will be some anomaly in treatment of people in different situations. This reflects the anomalies inherent in conventional measures of cash incomes. For the purposes described here, approach 4 appears to involve the least damaging inconsistencies particularly between those in different tenures, and is the easiest to calculate, so this is adopted in the With Housing Income measure. A case can, however, be made for the alternatives.

'Before' and 'After Housing Costs' measures

Even ignoring the problems of capital gains and mortgages, creating With Housing Income measures is not straightforward. In particular, it requires good estimates of the market rents which *would* be paid for the properties occupied by the majority of households covered by a survey, but who do not, in fact, pay a market rent. There are different ways of doing this (see Gardiner *et al.* 1995: section 4.2), but it is only straightforward to do so in countries which have a significant part of their housing stock both in the private rented sector and free of rent control. In the 1980s only a minority of member states of the European Union were in this position (see Tables 9.1 and 9.5).

One way to try to circumvent these problems has been applied in the UK, where official series are produced for income distribution both Before Housing Costs (BHC) and After Housing Costs (AHC) (see DSS 1993; Johnson and Webb 1990; Atkinson *et al.* 1993; Harris and Davies 1994). An advantage of the After Housing Costs measure is that it removes distortions between tenants with the same net housing costs, but subsidised through different routes (housing allowances as opposed to below-market rents). In the UK this has been important in making income distribution comparisons over time periods when there have been shifts from 'bricks and mortar' subsidies to income-related housing allowances, as there were over the 1980s. The same would be true in comparing two countries, one of which relied heavily on income-related housing allowances (conventionally included in measures of net income), while the other made greater use of below-market rents as a way of assisting those on low incomes.

Furthermore, individuals may, in fact, have little 'choice' in the housing they live in, particularly those who are allocated to social housing. In this situation, the rent they pay has some of the characteristics of a 'tax', rather than of a price

Table 9.1 Housing tenure (%)

	Year	Owner-occupiers	Private rented sector	Social rented sector	Other	Source
Spain	1989	88	11	1	—	Ghékiere (1991)
Ireland	1990	78	9	14	—	Power (1993)
Greece	1981	70	27	—	4	Ghékiere (1991)
UK	1992	66	10	24	—	DoE (1994)
Italy	1990	64	24	5	7	Ghékiere (1991)
Belgium	1986	62	30	6	2	Ghékiere (1991)
Luxembourg	1981	59	35	—	5	Ghékiere (1991)
Denmark	1991	58	16	21	5[a]	Power (1993)
Portugal	1981	56	36	4	4	Ghékiere (1991)
France	1992	54	21	17	8[b]	Lacroix (1994)
Netherlands	1991	46	12	41	—	Teule (1994)
Germany	1987	38	43	15	4	Hubert (1993)

Notes
a Private co-operatives.
b Of which 6.7 per cent are rent-free dwellings.

paid for a freely chosen amount of consumption. Treating rent as a tax suggests comparing income after housing costs, not before them.

The drawback of AHC measures is that they only give a fair comparison of relative living standards between households if they occupy accommodation of the same quality or value. The fact that a household has little in the way of net resources left over for other forms of consumption because it has chosen to spend most of its income on living in a luxury apartment in the centre of a capital city does not mean that it is appropriate to place it 'in poverty' by comparison with another household with slightly higher AHC income, but occupying a low cost hovel.

A further problem is that part of what is conventionally taken as the 'housing costs' of owners with a mortgage in effect represents the cost of acquiring an asset – that is, saving – rather than of consumption. This is obviously the case for the part of payments which represents 'principal repayment' rather than interest. But, as noted above, at times of inflation, the same will also be true of part of the nominal interest payments made. Someone who starts and ends a year with the same cash debt has improved their position in real terms. The nominal interest rate which they pay can be seen as equal to the real interest rate plus an amount giving 'compensation' to the lender for the way in which inflation erodes the real value of the debt. In real terms, therefore, part of the nominal interest paid represents principal repayment. Calculating the AHC income of mortgagors by deducting all of their nominal interest payments at a time of inflation does not take account of this phenomenon and will thus distort comparison between tenants and mortgagors (making the latter look unduly poor), and between mortgagors in countries with different inflation rates.

In the empirical results for the UK presented below we examine whether After Housing Costs measures give results closer to the theoretically preferable With Housing Income measure.

Relationships between different income measures

The relationships between these different income measures are shown in Figure 9.1 (for owner-occupiers) and Figure 9.2 (for tenants). In both cases the links are shown between net cash income before allowing for any effects of the housing system on the left of the diagram, and real comprehensive income on the right.

Owner-occupiers

Looking first at the position of owner-occupiers in Figure 9.1, several housing-related adjustments may already be incorporated into conventionally measured net cash income (Before Housing Costs). First, net cash income will be calculated after deducting direct taxes, and the size of these may be affected by tax reliefs for mortgage interest,[7] depreciation (as in Germany), or repair and maintenance costs. Second, net cash income will include any income-related housing

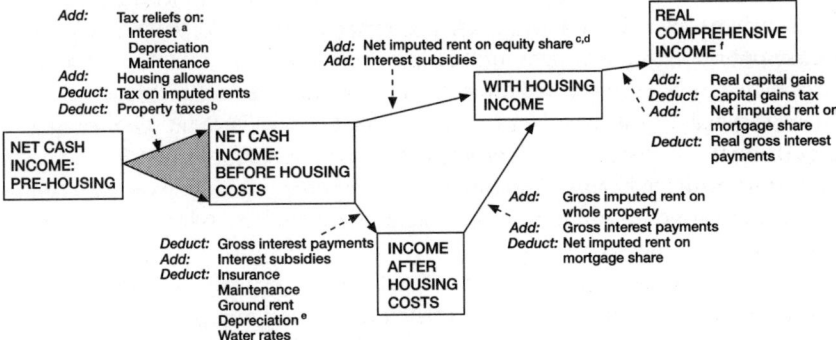

Figure 9.1 Components of housing income and costs: owner-occupiers

Notes

a UK 'Before Housing Costs' definitions treat interest relief as an interest subsidy, and do not deduct.

b Deducted in some definitions of net income (e.g. UK series), but not others (e.g. Eurostat 1990, Hagenaars *et al.* 1994).

c Size of equity share will reflect purchase grants, bonuses on housing savings schemes, purchase discounts, etc.

d Net imputed rent on equity equals: net imputed rent on whole property *minus* (net imputation rate/gross interest rate × gross interest payments).

e Not allowed for in UK official estimates.

f Real comprehensive income equals: income BHC *plus* net imputed rent on whole property *plus* real capital gains *minus* real net interest payments *minus* capital gains tax.

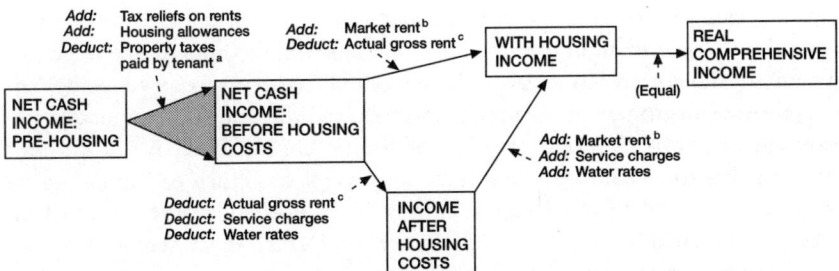

Figure 9.2 Components of housing income and costs: tenants

Notes

a Depending on definition of net income.

b Also add value of effective equity acquired if in 'rent to mortgage' scheme.

c Actual gross rent will reflect general subsidies (if passed to tenant), reductions for low income tenants, additions for high income tenants, underground rent additions, tax advantages of private landlords (if passed to tenant) and effects of rent control in private rented sector.

163

allowances received by owner-occupiers. It will also be net of direct tax charges, which may be related to housing, as when owner-occupiers are liable to tax on an assessment of the imputed rent on their property. On some definitions, property taxes may also be counted as a direct tax and be deducted at this stage, while in others they will be classed as a tax on expenditure and will not be.

In order to calculate the With Housing Income measure, two further additions have to be made: first, an addition for the net imputed rent on the owner's equity share; and second an addition for the value of any interest subsidies granted on borrowing by owner-occupiers.[8]

Subsidies to owners can also take the form of initial capital grants for purchase, discounts for social tenants to purchase their home, and bonuses or tax privileges on special housing-related savings schemes. The effect of all of these will be to increase the owner's equity share in the house, and so boost With Housing Income by increasing the amount on which net imputed rent is calculated.

By contrast, starting from BHC income, deductions for all of the main components of housing costs should be made to derive AHC income, including interest payments (less any subsidies or tax relief not already allowed for), insurance, repair and maintenance costs, ground rent and depreciation (although direct estimates of this are unlikely to be available from household survey data). In the UK water charges (which generally still depend on the size of property) are also counted as a housing cost in official calculations.

Tenants

The situation for tenants is rather simpler, as shown in Figure 9.2. Conventional measures of net income will generally incorporate the effects of any tax reliefs given for rent payments, of income-related housing allowances, and – depending on the definition – property taxes which are paid by the tenant or occupier.

To derive the With Housing Income measure requires addition of the difference between the actual gross rent paid and an estimate of the gross rent which would be charged in an unrestricted market. This will incorporate the effects of general subsidies going to social landlords, of reduced rents for low income tenants, the benefits to tenants from rent controls, and the value of accommodation provided free or at reduced rents by employers. On the other hand, if a rent surcharge has to be paid by high income tenants (as in French or German social housing), or if underground rent additions are paid in addition to 'controlled' rents (as happened in Italy, at least until recent reforms), the difference between actual and market rents will be diminished.

For tenants, the With Housing Income measure is the same as real comprehensive income. To derive AHC income for tenants simply requires the deduction of gross actual rents from BHC income, together with any other associated costs, like service and water charges.

Effects of tenure patterns

Looking at Figures 9.1 and 9.2, and comparing the adjustments required to go from conventionally measured net income to the preferable With Housing Income measure, it is evident that the scale of the adjustments will be affected by a country's tenure pattern. In particular, if tenure varies with income, the shape of the With Housing Income distribution may differ from that of conventional net cash income. As Table 9.1 shows, tenure patterns vary widely across member states, with the proportion of owner-occupiers varying from under 40 per cent in Germany to nearly 90 per cent in Spain, and the proportion in the social rented sector from virtually none in Greece and Luxembourg to over 40 per cent in the Netherlands.

The implication of Table 9.1 is that the difference between average cash income and With Housing Income is likely to be substantial in all member states, but to be more important in some than in others. In particular, it will be more important in countries with only small uncontrolled rented sectors (most member states) than in those like Germany where a substantial proportion of households pay a market rent.

Furthermore, tenure patterns vary in different ways across income groups. One indication of the scale of this effect is shown in Table 9.2, drawn from Hagenaars *et al.* (1994). The results shown are based on expenditure (rather than income) data *including* estimates of imputed rents but not deducting interest payments for owner-occupiers (and not including imputed additional housing expenditure for tenants paying below-market rents). The table shows the percentage of households within each tenure group which spend less than 50 per cent of overall average. It also shows the relative importance of each tenure group.

In the five countries where data are separately available, owners with mortgages emerge as having far smaller incidence of low expenditures than the other groups. Only in Greece do owner-occupiers as a whole have a higher incidence of low expenditures than do all households. In nine countries tenants as a whole have a higher incidence of low expenditures than the national average, with social tenants having particularly high rates in the UK and Ireland where they are separately identified. The residual 'other' category of households (often those occupying rent-free accommodation provided with a job) also have generally high incidence of low expenditures.

The treatment of owners with mortgages may have important effects on international comparisons for two reasons. First, in several countries mortgagors tend to be a relatively high income group. Adding full imputed rents without any adjustment for their mortgages may therefore *exaggerate* income inequality at the top of the distribution, while deducting full mortgage costs (as in calculating AHC income) when nominal interest rates exceed real ones may *understate* inequality at the top. Second, the level of indebtedness of owners varies between countries where they make high reliance on credit systems and others, particularly southern member states where credit systems are less developed and there is

Table 9.2 Households below half average expenditure by tenure

	% of households below 50% average expenditure[a]						Households in group as % of all				
	All	Outright owners	Owners with mortgage	Private tenants	Social tenants	Other[b]	Outright owners	Owners with mortgages	Private tenants	Social tenants	Other
Portugal (1989)	26.5	30.0	3.9	26.4		32.2	49.4	9.8	32.0		8.7
Italy (1988)	22.0	21.0		24.0		22.9	66.3		28.5		5.3
Greece (1988)	20.8	23.0		13.2		25.2	n.a.		n.a.		n.a.
UK (1988)	17.0	13.2	4.3	22.6	38.8	17.2	23.8	41.7	6.1	26.7	1.7
Ireland (1987)	16.4	13.2	7.6	15.5	46.9	28.6	41.3	35.4	7.8	13.9	1.6
Spain (1988)	16.3[c]	13.1		30.5		19.3	77.0		15.6		7.5
France (1989)	14.9	14.8	5.8	18.7		23.7	33.0	22.4	37.2		7.3
Germany (1988)	12.0	6.3		16.8	16.8	—	46.4		53.6		—
Luxembourg (1987)	9.2	7.0		16.0		7.3	72.2		24.5		3.2
Belgium (1987–1988)	6.6	4.0		12.3		5.4	66.1		30.5		3.3
Netherlands (1988)	6.2	0.3	0.5	10.8	10.8	0.0	8.2	35.8	55.3		0.7
Denmark (1987)	4.2	3.4		3.8		19.5	55.8		40.0		4.2

Source: Hagenaars et al. (1994), Appendix Tables A4.6 for each country (except A3.1 for Greece).

Notes

a Equivalent mean household expenditure using modified OECD equivalent scale.
b Including rent-free.
c Given as 17.5 in other tabulations.

more reliance on self-finance and even self-construction. In the latter countries, less of an adjustment to owners' imputed rents will result from taking account of mortgages.

Clearly then, adjustments which have differential effects between and within tenure groups will not only affect the shape of income or expenditure distributions, but will also do so in ways which vary between countries. This makes it potentially very important to allow for housing income in, for instance, making international comparisons of poverty rates.

Effects of tax and subsidy systems

Appendix 1 of Gardiner *et al.* (1995) describes the relevant features of the housing tax and subsidy systems of the member states.[9] Table 9.3 summarises the features affecting owner-occupiers, while Table 9.4 does the same for tenants. Looking back to Figures 9.1 and 9.2 it is evident that the means by which subsidies or tax concessions are granted will affect income under some definitions but not others. As a result, some income definitions may not provide a fair comparison between different countries, particularly those measured before adjustment for housing costs or housing income. For instance:

1   Tax reliefs and income-related housing allowances for owner-occupiers will generally affect measures of net cash income, but concessions to owners taking the form of interest subsidies or purchase grants of different kinds will not. Nine of the member states listed use interest subsidies to encourage owner-occupation (in most cases with the size of subsidy related to income). Omitting the advantage of these subsidies could distort comparisons between them and countries like Ireland where they are not used.

2   The way in which taxes are levied on owners also varies across countries. Eight countries include some kind of estimate (usually an underestimate) of imputed rents in the taxable income of owners, while most have some kind of property tax (local or central) affecting owners. While the income tax on imputed rents will generally be reflected in net incomes, property taxes may not be, again possibly distorting comparisons between countries placing more or less reliance on the different kinds of tax.

3   Turning to tenants, income-related housing allowances are generally included in net cash incomes, but the advantages of below-market rents for tenants are not. Failing to allow for such advantages will distort comparisons between countries which make more use of housing allowances (like Germany and, to some extent, the UK) and those which make more use of subsidies and rent controls (like the Netherlands with its large social rented sector or Portugal with its relatively large private sector subject to rent control).

4   Several countries use systems by which the rents charged to low income social tenants are reduced at source rather than explicit housing allowances. The

Table 9.3 Comparative treatment: owner-occupiers

	B	DK	F	G	GR	IR	IT	LUX	NL	P	S	UK
Taxation												
Tax reliefs:												
– interest	✓[1]	✓[3]	✓[7]	✓[9]	✓[12]	✓[14]	✓[17]	✓[19]	✓	✓[19]	✓[19]	✓[25]
– depreciation	✓[2]			✓[10]							✓	
– maintenance	✓[2]	✓[4]	✓[8]									
Tax on imputed rents	✓	✓			✓[13]		✓	✓	✓[20]	✓	✓	
Property taxes	✓	✓	✓	✓[11]	✓			✓[18]	✓	✓[21]	✓	✓[26]
Capital gains tax		✓[5]			✓[5]					✓[22]	✓[24]	
Subsidies												
Public grants:												
– general	✓					✓[15]		✓		✓[23]		
– social tenants	✓					✓	✓			✓		✓
Bonuses on housing savings schemes		✓	✓	✓	✓			✓		✓		
Interest subsidies:												
– general		✓			✓				✓			
– income-related	✓		✓		✓		✓	✓	✓	✓	✓	
Housing allowances		✓[6]	✓	✓		✓[16]						✓[27]

Notes
1 Up to limit of taxable imputed rent.
2 Fixed 40 per cent deduction from imputed rent for maintenance and depreciation.
3 50 per cent tax credit since 1987.
4 Deduction of 1 per cent of taxable value (within limits) against tax on imputed rents.
5 If sold within two years of occupation.
6 Retired households only.
7 Tax credit at 25 per cent for five years if income below limit.
8 Tax credit for certain renovation work.
9 Before 1987 only, plus temporary relief for new construction 1991–1994.
10 Depreciation allowance for new construction of (since 1987) 6 per cent for four years and 4 per cent for four years (deduction from taxable income).
11 Ten-year exemption for new construction.
12 Limited to 50 per cent of declared income, except for first-time buyers.
13 Since 1988 only applies to property over 200 square metres.
14 Since 1989 only 80 per cent of interest deductible (up to limit).
15 First-time buyers.
16 Interest payments for unemployed and others with low income via Supplementary Welfare Allowances.
17 Tax credit of 27 per cent up to limit depending on date of purchase.
18 Tax on rental value replaced by central taxes on capital value from 1992, with local tax on surface area from 1994.
19 Up to limit.
20 Tax rising from 1.8 per cent (1990) to 2.8 per cent (1994) of occupied value (60 per cent of unoccupied value).
21 With exemptions.
22 On half of capital gains adjusted for inflation.
23 For renovation of older property.
24 With roll-over relief if reinvested in a new property within two years.
25 Up to limit on amount of loan eligible for relief. Relief now given at fixed rate (25 per cent up to 1993/94, 20 per cent in 1994/95, 15 per cent since 1995/96) via lenders. Additional relief for higher rates of income tax via tax system up to 1990/91).
26 Did not apply between 1990/91 and 1992/93.
27 Owner-occupiers receiving Income Support can have part or all of mortgage interest paid for them.

Table 9.4 Comparative treatment: tenants

	B	DK	F	G	GR	IR	IT	LUX	NL	P	S	UK
Taxation												
Tax relief on rents					✓[8]	✓[9]					✓[17]	
Housing allowances												
Social tenants		✓	✓[3]	✓					✓			✓
Private tenants	✓[1]	✓	✓[3]	✓		✓[10]			✓			✓
Social tenants												
General subsidies	✓	✓	✓	✓		✓	✓		✓[14]	✓	✓	✓
Reduced rents for low incomes	✓					✓	✓	✓		✓	✓	
Higher rents for high incomes			✓	✓								
Rent-to-mortgage scheme	✓											✓
Private rented sector												
Extensive rent control		✓			✓		✓[12]	✓	✓[15]	✓	✓[18]	✓[19]
Underground rent additions					✓		✓[12]					
Residual rent control			✓[4]	✓[6]		✓	✓[13]					
Constraints on rent increases			✓	✓								
Subsidies to landlords			✓[5]						✓[16]			
Tax concessions to landlords	✓[2]			✓		✓[11]						✓[20]
Property taxes	✓	✓	✓	✓[7]	✓		✓		✓	✓	✓	✓[21]

Notes

1 For displaced households and elderly tenants.
2 Reduced tax on rent receipts if tenant household has more than two children.
3 Type of allowance and eligibility depends on type of loan used for construction/renovation, and on quality of dwelling.
4 About 12 per cent of sector still subject to 'Law of 1948' rent controls.
5 For renovation.
6 In Berlin only (until end of 1993).
7 With-ten year exemption for new construction.
8 Salaried employees and pensioners can deduct 30 per cent of rent from taxable income up to limit.
9 For tenants aged fifty-five or older if not eligible for Supplementary Welfare Allowances.
10 For tenants affected by rent rises caused by decontrol after 1981.
11 For new-build properties since 1984.
12 Up to 1992 extensive fair rent control system, but accompanied by undeclared rent additions.
13 New lettings since 1992 freely negotiated, but with security for fixed term.
14 Since 1995 cash flow subsidies to social landlords.
15 Control does not apply to property above certain value (and unofficial low quality sector operates outside controls).
16 For providers of new housing.
17 Tax credit of 15 per cent of rent up to limit for low income tenants.
18 Liberalisation of rents on new tenancies since 1985.
19 Effective decontrol of rents on new tenancies since 1988.
20 Substantial tax concessions for a limited amount of new investment under the Business Expansion Scheme between 1988 and 1993.
21 Did not apply between 1990/91 and 1992/93.

advantages of such rent reductions generally will not be reflected in income measures (although they would be in AHC incomes), but housing allowances will be. Again, this will seriously affect comparisons between countries unless they are taken into account.

Overall, the survey of housing systems in Tables 9.3 and 9.4 shows that there are major differences between member states in the ways in which housing is subsidised or taxed, and that these differences will distort comparisons between tenures, income groups and countries if they are made on the basis of conventionally measured cash incomes, even if those incomes include an allowance for the imputed rents of owner-occupiers. A With Housing Income measure avoids these problems. Some of the problems are also avoided by using AHC income, but only at the cost of introducing the additional problems described above.

Examples for France and the United Kingdom

To illustrate the potential scale of the effect of allowing for housing income in comparing distributions, we constructed estimates of the distribution of housing income in France (1984–1985) and the UK (1989), using microdata from each country's household budget survey (the EBF [Enquête Budget de Famille] in France and FES [Family Expenditure Survey] in the UK).[10] The results are summarised in Table 9.5, which shows the effects on two aspects of income distribution: the *numbers* counted as falling below various relative 'poverty lines' in its top panel, and the income shortfall from each line – the *poverty gap* – in the lower panel.

Distribution before adjustment

The starting point is the distribution of cash income, *including* explicit housing allowances (Housing Benefit in the UK, benefits like APL in France), but without adjustment for other forms of housing income or costs. On this definition, poverty was significantly higher in the UK in 1989 than in France in 1984–1985. Taking the 50 per cent of mean income threshold, 17.1 per cent of those in the UK fell below it, as opposed to 12.5 per cent of those in France. However, at the lower threshold, 40 per cent of mean income, the difference was narrower: 8.2 per cent being below it in the UK and 6.4 per cent in France.

This reflects the shape of the distribution of the low income population in the two countries, further exemplified in the lower panel of Table 9.5. This gives the 'poverty gaps' under different definitions, that is the total income shortfall of those who fall below a threshold, expressed as a percentage of total household income. At the 50 per cent threshold, the shortfall is 1.9 per cent in the UK, only slightly higher than the 1.8 per cent in France. Substantially more individuals fall below the threshold in the UK, but they do not fall so far below it, so the aggregate income shortfall is only slightly greater. This is made clear in looking at

Table 9.5 Estimated proportions of individuals with low incomes and poverty gaps in the United Kingdom (1989) and France (1984–1985)

(a) *Numbers with low incomes*

Income definition	United Kingdom			France		
	Proportion (%) less than:			*Proportion (%) less than:*		
	40%	*50%*	*60%*	*40%*	*50%*	*60%*
Base	8.2	17.1	27.1	6.4	12.5	22.0
Y1 With Housing Income (rental method)	—	—	—	5.8	11.6	20.9
Y2 With Housing Income (capital value method)	7.8	15.7	25.6	5.5	11.2	20.6
Y3 After Housing Costs	12.5	22.7	31.2	—	—	—

(b) *Poverty gaps*

Income definition	United Kingdom			France		
	Poverty gap (%) with threshold as % of mean income:			*Poverty gap (%) with threshold as % of mean income:*		
	40%	*50%*	*60%*	*40%*	*50%*	*60%*
Base	0.74	1.90	4.01	0.91	1.80	3.46
Y1 With Housing Income (rental method)	—	—	—	0.74	1.54	3.08
Y2 With Housing Income (capital value method)	0.70	1.75	3.71	0.66	1.42	2.90
Y3 After Housing Costs	1.72	3.39	6.03	—	—	—

Notes
In UK estimates, housing income of mortgagors is calculated on the basis of their equity shares. In the estimates for France, housing income for mortgagors is unadjusted for mortgages. The two sets of adjustments are therefore not fully comparable.

the poverty gaps at the 40 per cent level. In this case the aggregate shortfall in the UK, 0.74 per cent, is actually smaller than that in France, 0.9 per cent, despite there being a bigger proportion of the population below the threshold in the UK.

By themselves, these results emphasise that the scale of numbers counted as having low incomes, and even the ordering between countries, can be sensitive to the threshold chosen, and that a single measure – numbers below one threshold – fails to capture important features of differences in income distribution.

Allowing for housing income

The next three rows of each panel show the effects of adjusting for housing in different ways. Because the free private rented sector in the UK is so small, estimates of housing income based on data from rents in the private sector (the 'rental value' method) cannot be constructed, but this can be done with the French data (using hedonic rent indices derived from data within the EBF). Instead, housing income is calculated using assumed rates of return and costs applied to estimated capital values (using hedonic price indices drawn from sales data, the 'capital value' method). Given that UK owner-occupiers tend to have higher incomes than tenants, one might have expected that in the UK, the With Housing Income measure would be distributed less equally than under the base definition. In fact, our results suggest that allowing for housing income (on the capital value method) very slightly *narrows* the income distribution in the UK, reducing the numbers below each of the thresholds and the poverty gaps, but not greatly. This turns out to be very much the same pattern as in France using either of the methods.[11]

These results may at first sight seem surprising to those who expected a larger effect from the inclusion of the imputed rents of owner-occupiers in terms of raising the incomes of those in the higher income groups. However, what turns out to be important is the fact that we are also including – which conventional adjustments for housing usually do not – the benefits to low income tenants from paying below-market rents. These amounts may be smaller in absolute terms, but they can be as important *relative to their cash incomes*. In addition, it should be borne in mind that it is the way in which cash and housing incomes are related to each other which will matter, rather than the inequality of housing income *per se* (see, for instance, Lerman and Lerman 1986).

This can be seen in the UK case from the top and bottom panels of Table 9.6. This shows, broken down by tenure type, income under various measures for those in different quintile groups (fifths) of the population (ordered by income under the base definition). The top panel shows average income under the base definition, while the bottom panel shows 'housing income'. As one might expect, the largest averages for those in all income groups are for those living rent free and for owner-occupiers owning outright. These are followed by owners with mortgages. Their imputed rent is not as great as for the outright owners (reflecting the fact that they have only partial equity stakes on which to be imputed this income), but they benefit from the mortgage interest relief system. Their housing income rises faster with income than for the outright owners. Meanwhile, the housing income of the tenants living in social housing (the first two categories) or benefiting from rent control (most of those in the third column) is smaller.[12] However, these tenants are concentrated in the bottom two quintile groups, where the amounts involved can represent over 10 per cent of their base income. By contrast, owners with mortgages are concentrated in the higher income groups. For instance, for owners with mortgages in the top group, housing income only represents 5 per cent of their base income.

Table 9.6 Income[a] before and after housing costs, and housing income by tenure in the United Kingdom (1989)

Quintile group (by income before housing costs)	Local authority	Housing association	Other rented un-furnished	Other rented furnished	Owners with mortgages	Outright owners	Rent free
(a) *Income before housing costs*							
Bottom	46.81	48.10	41.64	39.74	43.23	48.81	41.23
2	71.28	71.41	70.50	69.80	74.20	72.33	73.58
3	96.65	96.47	94.82	96.88	98.48	98.18	97.71
4	129.46	132.42	135.26	133.24	133.20	134.05	134.43
5	188.91	195.49	256.52	227.22	231.70	233.91	190.23
All	76.03	76.09	90.89	111.94	139.12	116.83	89.36
(b) *Income after housing costs*							
Bottom	34.28	35.64	26.45	15.25	26.78	41.44	34.84
2	53.09	52.26	52.62	42.65	56.93	64.74	68.04
3	81.22	74.60	76.48	66.46	78.93	91.03	91.60
4	112.96	111.25	120.32	103.35	110.02	125.79	128.94
5	169.92	170.10	230.57	184.39	191.15	223.30	175.90
All	60.71	58.68	73.59	81.04	112.91	108.66	82.86
(c) *Housing income*							
Bottom	4.34	4.56	6.18	−5.13	6.85	11.92	21.55
2	5.07	4.50	8.07	−7.77	7.75	12.53	20.03
3	5.50	4.60	10.44	−6.32	8.51	12.65	27.85
4	7.18	8.21	12.11	−3.40	9.69	13.36	20.10
5	9.71	16.79	2.73	−10.61	12.24	16.34	35.49
All	5.23	5.37	7.06	−6.78	9.68	13.34	23.05

Note
a   Income is equivalent income per individual.

This suggests an important general conclusion: when adjusting for housing income it may be misleading to adjust *only* for the imputed rents of owner-occupiers and those living rent free – which is exactly what some countries' income distribution statistics do. Doing so may produce a less equal distribution, and hence raise the numbers counted as having low incomes. However, the benefits of low rent accommodation could actually be greater in relative terms for those on low incomes, so that the overall effect would be a narrowing of the distribution.[13]

Income After Housing Costs

Table 9.5 also shows the numbers counted as having low incomes under the After Housing Costs definition presented in official statistics in the UK. The effect of this is to produce a substantially less equal distribution, raising the numbers with incomes below half of the mean from 17.1 to 22.7 per cent, for instance. Looking at Table 9.6, part of the reason for this can be seen to be the very large reduction in income for low income mortgagors going from the top to second panels. Low income mortgagors have only small amounts of income to spend on other purposes, but they are both purchasing significant amounts of housing consumption, and are making an investment, so deducting all their mortgage costs misrepresents their long run position, particularly if the inflation rate is high.

This suggests a more general conclusion. Although there are circumstances (like a cut in subsidies to social tenants) in which the AHC measure may give a better guide to trends in living standards over time, in this case at least, the AHC measure not only produces a larger effect than the inclusion of housing income, but it may actually affect the results in the wrong direction. The base, BHC, measure appears to be a better approximation to our preferred, With Housing Income, measure than the AHC measure.

Conclusions

This review suggests not only that there are strong arguments in principle for allowing for 'housing income' when comparing income distributions between countries, but also that doing so will have significant effects on the shape of income distributions, for instance in terms of the numbers recorded as, for instance, having incomes below half the average. The strength and direction of these will vary between member states. The discussion suggests that adjustment should be made not only for the imputed rents of owner-occupiers (taking account of mortgages), but also for other flows of housing income, including the advantages of paying below-market rents for social tenants and private tenants affected by rent control, and various forms of interest subsidy for owners with mortgages.

Allowing for housing income is not simply a matter of allowing for the imputed rents of owner-occupiers and those living rent free. It is also important to allow for the benefits in kind accruing to those paying below-market rents. In making such adjustments, our preferred methodology would be to produce estimates of the *net* imputed rent of owner-occupiers, accruing on their equity shares, that is, only a percentage of the net imputed rent for those with mortgages (although other approaches are possible). In countries with a large, uncontrolled private rented sector it should be possible to derive estimates of imputed rents from data on *gross* rents in the private sector, and to make a percentage deduction for costs like repair, maintenance, depreciation and insurance to arrive at a net imputed rent. For mortgagors a further deduction should be made representing the net imputed rent on the part of capital value covered by the mortgage (but *not* by deducting actual nominal mortgage payments). However, for the majority of member states, the uncontrolled private rented sector has been too small in recent years to produce reliable estimates using the rental method. In these countries, better estimates can be made of capital values, and net imputed rents calculated as a percentage of these (again, of equity shares for those with mortgages). For mortgagors, housing income should also include the value of subsidies which they receive on mortgage payments.

For tenants paying anything other than an uncontrolled market rent, housing income should include an estimate of the difference between the actual gross rent and an estimated gross market rent (using the same methodology as for owners). In some cases this may be negative. Using data from the private sector for either rents or capital values to derive these estimates may be inaccurate, if there are special environmental factors affecting social housing, for instance, which cannot be captured by the characteristics used in the derivation of hedonic rent or capital value indices.

Both in principle and in practice, the *ad hoc* adjustment of calculating incomes After Housing Costs appears to be an unsatisfactory way of allowing for differences between tenures. Making no adjustment for housing at all seems to produce a better approximation to our preferred With Housing Income measure than does deducting housing costs.

Finally, the way in which some of our empirical findings did not match a priori expectations suggests that it is hard to predict the effects of these adjustments without actually making them for each country. The shape of income distributions and the numbers with relatively low incomes will be affected by the size of adjustment for each household in proportion to their incomes; these are hard to predict without use of microdata. A considerable research task remains therefore, to construct consistent and internationally comparable estimates of the distribution of housing incomes, and of the factors like subsidies and tax concessions which contribute to it. This would not only improve our knowledge of income distribution differences between countries, but would also give us comparisons of the distributional effectiveness of different kinds of housing subsidy system.

Notes

1 This paper is drawn from joint work with Karen Gardiner and Jane Falkingham of the LSE Welfare State Programme, Valérie Lechene of the Institut National de Recherche Agronomique in Paris, and Holly Sutherland of the Microsimulation Unit, Cambridge University. In particular, the empirical results presented here were calculated by Karen Gardiner. The work was originally financed and carried out for Eurostat, and full details can be found in Gardiner *et al.* (1995). The work was carried out before the recent enlargement of the EU, so the information presented relates to twelve member states only. Anonymised Family Expenditure Survey data used in the study were kindly supplied by the Central Statistical Office and ESRC data archive at Essex University; interpretation of the data is the responsibility of the author. The author is grateful to the editors and to an anonymous referee for pointing out inconsistencies in an earlier draft.

2 See Yates (1994) for a related discussion of issues raised by UN recommendations for including imputed rents of owner-occupiers in national income and distributional estimates.

3 The true value to the tenant of this 'advantage' will be reduced if, given the same cash sum, the tenant would have chosen to spend it in some other way, rather than on the particular accommodation to which the advantage is tied. In practice, making an adjustment for this is difficult. The problem should be borne in mind in considering the empirical results, particularly in respect of subsidies to tenants of social landlords.

4 Even this definition is not without conceptual problems, for instance when a change in asset values is caused by a change in real interest rates (see Meade Committee 1978: 31; Hills 1991: 190).

5 As with subsidised tenants, this may *over*state the advantage to owners. Given the cost of moving, long-standing owners may value their accommodation less highly than the market without it being worth their while to move. An alternative, lower bound, estimate could be obtained by calculating how much cash income owners are foregoing by not converting their equity share into a 'reverse' mortgage (see Venti and Wise 1991 for a calculation of this kind for a different purpose). Capital market imperfections mean, however, that this *under*states the value to owners.

6 See Yates (1994) for a discussion of this and associated problems.

7 In UK official statistics (as in the 'Households Below Average Incomes' series) BHC incomes are calculated before allowing for the advantages of mortgage interest tax relief. We follow this convention in calculating our 'base income' BHC for the UK and France. We allow for tax relief by deducting net interest payments, after relief, when calculating AHC income. However, we cannot identify separately the small amount of additional tax relief granted (until 1991) to higher rate income taxpayers in excess of the basic rate of relief given to all mortgagors, so our base income is, in fact, net of this element.

8 The definition of comprehensive income implies that to transform this measure into real comprehensive income would require an addition for real capital gains (net of any resultant capital gains tax) and for the rest of the net imputed rent (on the mortgage share) less the real gross interest payments made.

9 See also Ghékiere (1991), Maclennan and Williams (1990), Boelhouwer and van der Heijden (1992), Papa (1992) and van Vliet (1990); for a full list of sources, see Appendix 1 to Gardiner *et al.* (1995).

10 The methods used, and the substantial practical problems encountered, are described in Gardiner *et al.* (1995).

11 The two sets of estimates are not fully comparable. For technical reasons resulting from limitations in the data to which we had access, we were not able to allow for mortgages in the French results (either for their effect in reducing the equity share on which owners' imputed rents should be based, or for the value of subsidised mortgages available through various schemes). This will have led to an overstatement of the housing income of owners with mortgages and thus probably to an overestimate of the numbers with low incomes relative to the mean.

12 Income from housing allowances – Housing Benefit – is already included in the base definition.

13 This is subject to two qualifications. First, the characteristics used to determine estimated market rents for social tenants may omit important environmental factors which would in reality depress market rents. Second, as discussed above, a rationed benefit may not be of as great value to its recipient as its cost to the provider. Both of these suggest that the housing incomes shown, for instance, in Table 9.6 for social tenants may be exaggerated.

References

Atkinson, A.B., Gardiner, K., Lechene, V. and Sutherland, H. (1993) *Comparing Poverty in France and the United Kingdom*, Welfare State Programme Discussion Paper WSP/84. London: London School of Economics.

Boelhouwer, P. and van der Heijden, H. (1992) *Housing Systems in Europe, Part I: A Comparative Study of Housing Policy*. Delft: Delft University Press.

DoE (Department of the Environment) (1994) *Housing and Construction Statistics, December 1993, Part 2*. London: HMSO.

DSS (Department of Social Security) (1993) *Households Below Average Income 1979–1990/91*. London: HMSO.

Eurostat (1990) *Poverty in Figures: Europe in the Early 1980s*. Luxembourg: OOPEC.

Gardiner, K., Hills, J., Falkingham, J., Lechene, V. and Sutherland, H. (1995) *The Effects of Differences in Housing and Health Care Systems on International Comparisons of Income Distribution*, STICERD Welfare State Programme Discussion Paper WSP/110. London: London School of Economics.

Ghékiere, L. (1991) *Marchés et Politiques du Logement dans la CEE*. Paris: Documentation Française.

Goode, R. (1980) 'The Superiority of the Income Tax', in J.A. Pechman (ed.) *What Should Be Taxed: Income or Expenditure?* Washington, DC: Brookings Institution.

Hagenaars, A.J.M., de Vos, K. and Zaidi, M.A. (1994) *Living Conditions of the Least Privileged in the European Community*. Luxembourg: Eurostat (mimeo).

Harris, G. and Davies, M. (1994) *Income Measures for Official Low Income Statistics: The Treatment of Housing Costs and Local Government Taxes*, Analytical Notes No. 2. London: DSS.

Hills, J. (1991) *Unravelling Housing Finance: Subsidies, Benefits and Taxation*. Oxford: Clarendon Press.

Hubert, F. (1993) *Germany's Housing Policy at the Crossroads*. Berlin: Institut für Wirtschaftspolitik und Wirtschaftsgeschichte, Freie Universität Berlin.

Johnson, P. and Webb, S. (1990) *Poverty in Official Statistics: Two Reports*, IFS Commentary No. 24. London: Institute for Fiscal Studies.

Lacroix, T. (1994) 'Tassement de la propriété et redressement du locatif privé', *INSEE–Première* 313 (May).

Lerman, D.L. and Lerman, R.I. (1986) 'Imputed Income from Owner-Occupied Housing and Income Inequality', *Urban Studies* 23: 323–331.

Maclennan, D. and Williams, R. (eds) (1990) *Affordable Housing in Europe*. York: Joseph Rowntree Foundation.

Meade Committee (1978) *The Structure and Reform of Direct Taxation*. London: Allen and Unwin.

Papa, O. (1992) *Housing Systems in Europe, Part II: A Comparative Study of Housing Finance*. Delft: Delft University Press.

Pechman, J.A. (ed.) (1977) *Comprehensive Income Taxation*. Washington, DC: Brookings Institution.

Power, A. (1993) *Hovels to High Rise: State Housing in Europe since 1850*. London: Routledge.

Teule, R. (1994) 'Income Support and Housing Subsidies in the Netherlands', York: Joseph Rowntree Foundation (mimeo).

van Vliet, W. (ed.) (1990) *International Handbook of Housing Policies and Practices*. Westport, CT: Greenwood Press.

Venti, S.F. and Wise, D.A. (1991) 'Aging and the Income Value of Housing Wealth', *Journal of Public Economics* 44, 3: 371–397.

Yates, J. (1994) 'Imputed Rent and Income Distribution', *Review of Income and Wealth* 40, 1: 43–66.

10

PAYING FOR OWNER-OCCUPIED HOUSING

Marietta Haffner

Comparisons of housing expenditure have been made in several countries, including the Netherlands, United Kingdom and France (Ministerie van VROM 1989; Department of the Environment 1990; Hills 1991; Taffin 1991). Such surveys usually use the household housing-expenditure ratio which shows the proportion of household income that is spent on housing. This focus on the expenditure of households can easily be explained by the way surveys, such as the British Family Expenditure Survey, the Dutch Housing Need Survey and the French Housing Survey, are carried out. The registration of outlay expenses (cash flows) induces expenditure comparisons. However, as Hills states: 'cash flow measures may be misleading as a guide to the true value of the advantages of owners and tenants' (1991: 28).

According to Van Order and Villani (1982: 87) cash flow measures had been used in the United States as well. However, in the last two decades American researchers have been utilising the concept of user costs of (housing) capital. It is a price concept for the use of capital and allows like-with-like comparisons among owner-occupiers, for example.

In this chapter these different means of measuring the affordability of housing are discussed in relation to owner-occupiers in six countries in the northwest of Europe. The meaning of these concepts of affordability will be analysed and their relative merits assessed. Both concepts will be applied to the financial situation of owner-occupiers in three countries.

The chapter is structured as follows. First, the financial instruments (taxation and non-fiscal instruments) for owner-occupiers in six northwestern European countries are described: Belgium, Denmark, France, the Netherlands, England (England rather than the United Kingdom is described, because of differences in the taxation systems in the constituent parts of the UK) and former West Germany. A discussion follows about the ways in which the financial treatment of owner-occupiers can be quantified. This discussion portrays the strengths and weaknesses of the concepts of housing expenditure and user costs. Calculations for sample households with different income levels in the Netherlands, Denmark and England illustrate what can be achieved with both concepts. It appears that the concept of expenditure is useful in certain ways, but that user costs must be used for an equivalent cost comparison to take place. The chapter concludes with a short discussion about how comparisons across countries could increase insight into the financial position of owner-occupiers by not only looking at expenditures and costs, but also at how to estimate the level of subsidisation.

Taxation and subsidies

The point of purchase, the period of occupation and the point of disposal of a dwelling form the impact points for taxation and subsidies for owner-occupied dwellings. However, disposal of a dwelling will be disregarded in the following discussion because the focus is on expenditure and cost during occupancy. The year 1990 is the base year so changes in taxes and subsidies after 1990 are not considered here.

Fees at the point of purchase

Table 10.1 shows the fees for the buyer of a dwelling at the point of purchase in the countries under study. A distinction is made between taxation and other fees. The Sixth Directive (1977) of the Council of Ministers (European Union) is applicable here specifying that newly built residences are subject to VAT.

Exceptions are England which applies a zero-rate and Germany and Denmark which exempt real estate transactions. The exemption clause induces contractors to include paid VAT in the selling price, as it cannot be reclaimed. Contrary to the situation in England, the Danish and German buyers thus effectively pay VAT as part of the selling price.

Table 10.1 Levies at the point of purchase paid by the buyer of an owner-occupied dwelling (percentage of purchase price, 1990)

Country	Existing dwelling		New dwelling	
	Total, range	Tax	Total, range	Tax
Belgium	16.1–18.2	12.5	21.4–23.5	17.0[a]
Denmark	2.0–2.4	1.2	2.0–2.4	1.2[b]
France	9.0–10.0	6.8[a]	22.0–23.0	18.6
Netherlands	7.7–9.4	6.0	20.4–22.4	18.5
England	1.0–1.6	1.0[c]	1.0–1.6	1.0[a,c]
(West) Germany	3.5–3.6	2.0	3.5–3.6	2.0[b]

Source: Own calculations, Haffner (1993, 1998).

Notes
a Includes subsidy for owner-occupied dwelling. Belgian VAT rate: 19.0%. French tariff: 15.4%. English VAT rate: 15.0%.
b Formal VAT exemption. VAT paid by contractor is included in price, however.
c Only levied when price is higher than £30,000.

Other acquisition costs complete the financial picture. These include registration and transaction taxes, and fees charged by lenders and lawyers.

Fiscal treatment during occupancy

Table 10.2 lists the most important ways in which taxation affects owner-occupiers after they have purchased a dwelling. The tax treatment of owner-occupiers in 1990 has not been subject to legislation by the European Union and falls into three categories: income, property and net wealth tax. The latter, however, is excluded from discussion. Capital gains tax which could be relevant during occupancy (accruals) is excluded as well as it is at best relevant when gains are realised at the disposal of a dwelling.

Unless otherwise noted, the general structure of taxation in 1990 is still relevant in 1996, though rates etc. have changed since. For instance, the English income tax rate for mortgage interest relief now is the lowest tax rate instead of marginal tax rate as it was in 1990 (De Kam 1997).

Owner-occupiers may be confronted with three types of income tax regulations.

Table 10.2 Relevant kinds of taxation[a] during occupancy for owner-occupiers (1990)

Country	Income tax							Property tax
	Rental value tax	Interest relief[b]	Concession for depreciation		Concession for maintenance			
			Actual	Lump sum	Actual	Lump sum		
Belgium	+[c]	+[c]	+[c,d]	+	−	+		−
Denmark	+	+	−	−	−	+		+[c]
France	−	+[c]	−	−	+[c,f]	−		+[g]
England	−	+	−	−	−	−		−[i]
Netherlands	+[h]	+	−	+	+[f]	+		+[g]
(West) Germany	−	−	+[c]	−	−	−		+[c]

Source: Haffner (1992b, 1994).

Notes
a A distinction between municipal and national taxes has not been made.
b Only in the Netherlands and Denmark is the whole amount of interest paid deductible.
c Amount is influenced by number of children.
d It concerns a limited relief for mortgage repayment.
e Tax is charged to the owner, not the user.
f It applies to extensive maintenance only. This concession was withdrawn after 1990.
g Tax is charged to both the owner and the user of real estate.
h Property tax, insurance premiums (and depreciation) are deducted via a lump sum from the rental value.
i A tax is charged per person older than 18 years, which replaced a property tax (rates) in 1990 and which has again been replaced by a (partial) property tax in 1993.

1   Deduction of mortgage interest payments (Belgium, Denmark, France, the Netherlands, England). The deduction is unlimited in the Netherlands and Denmark but various limits are applied in the other countries. These limits sometimes vary according to the number of children and the type of dwelling (new versus stock).
2   Tax on imputed rent (Belgium, the Netherlands and Denmark). Imputed rent or the rental value of the property to the occupier is determined as a percentage of dwelling value. In the Netherlands in 1990 imputed rent is calculated as a percentage of a banded value with the result that the higher the value of the dwelling within a class, the lower the percentage paid.
3   Other deductible items (all countries, except England). These items include concessions for maintenance, depreciation or repayment of capital. They are either available as a lump sum, for instance in the process of calculating rental value, or as deductions of actual amounts from taxable income or tax liability. In the case of actual amounts, limits are usually set.

   In Belgium a deduction from taxable income connected to the repayment of the loan exists. In Germany the depreciation allowance is not linked to debt finance. A percentage of the property value may be offset against tax. There is an upper limit to the property value, the allowance is time-limited and it is influenced by the number of children.

These allowances may be offset against the owner's marginal tax rate in all countries other than Denmark.

Property taxes are levied in all countries other than in Belgium, and in England where, between 1990 and 1993, a poll tax called the 'Community Charge' was levied on each adult. The other countries link the level of their (municipal) property tax to the property value: national and/or local rates are applied to the imputed rental value or to the (estimated) market value of land or the combined value of the land and the building.

Non-fiscal financial instruments

Table 10.3 shows the instruments that are intended to increase accessibility or affordability of owner-occupied dwellings in 1990. They usually are income-related, as low-income groups form the main target.

Through the contractual savings facilities Danish, French and German governments grant premiums for savings made on behalf of housing before a purchase takes place. A second advantage in Germany and France is the entitlement to a below-market-interest loan.

The one-off premium is paid in cash at the point of purchase in both the Netherlands and Belgium. Table 10.3 refers to the discounts made when English local authority tenants exercise their right to buy.

The recurrent contributions can be grouped into two classes: 'contribution' and 'below-market-interest loan'. In Germany a low-interest loan can signify

Table 10.3 Non-fiscal financial assistance for owner-occupiers (1990)

Country	Savings facility	One-off contribution	Recurrent contribution or rebate	
			Contribution	Low-interest loan
Belgium	–	+	–	+
Denmark	+	–	–	–
France	+	–	+[c]	+[c]
England	–	+[b]	–	–
Netherlands	–	+[a]	+[a]	+/–[d]
(West) Germany	+	–	–	+[e]

Source: Haffner (1992a, 1992b, 1994).

Notes
a   Newly built dwellings only.
b   Right to buy.
c   Only in combination with extensive renovation for existing dwellings.
d   Side-effect of guarantee: below-market-interest rate of about 0.2%.
e   Mainly newly built dwellings.

diverse concessions: for example, a postponement of the interest payments, a below-market rate and/or a repayment postponement.

The recurrent and one-off contributions in the Netherlands were abolished after 1990. However, the right to buy in England and the savings facilities in Denmark, Germany and France are still available.

Housing expenditure versus cost

The difference between the concept of expenditure and cost can best be demonstrated by a simplified example. In a certain year a dwelling is bought on 1 January for a price of 100 and sold on 31 December for 110. The acquisition is completely financed by a mortgage. There are no taxes and subsidies.

Annual expenditure or cash flow would relate to mortgage costs being interest charges plus capital repayment in case of a repayment mortgage. The increased owner's equity of 10 at the end of the year would be regarded as a cash inflow.

Annual (economic) cost on the other hand would embody other components. Repayment is not considered a cost, as the mortgage would be totally repaid at the sale. Interest payments would not be 'restituted' by the sale and would be regarded as a cost. Value increase, taking account of depreciation (value decrease because of ageing, for instance), would decrease cost.

Expenditure

Housing expenditure during occupancy consists of nominal net cash outflows in relation to the dwelling, i.e. the interest charges and repayment of capital in the case of a repayment mortgage, or contributions towards some other repayment vehicle, such as an endowment policy. The effects of the fiscal treatment of the owner-occupied dwelling (if relevant in a country) should also be taken into account: the deduction of mortgage interest and other items and the taxation of imputed rent, as well as property tax. Equation 1 presents all these components for the Dutch situation (Table 10.2). It incorporates the recurrent subsidy which is taxed as income (Table 10.3). As either a one-off contribution or a recurrent contribution can be received by the owner-occupier (Table 10.3), the one-off one is excluded from equation 1. The mortgage interest rate could either be at the full-market rate or at a below-market rate.

$$EXP_t = (1\tau)^*r^*D_{t1} + P_t + \tau^*RV_t + OZB_t(1\tau)^*S_t \tag{1}$$

where   EXP = housing outlays of the owner-occupier
          t   = year
          τ   = income tax rate
          r   = mortgage interest rate
          D   = debt
          P   = principal repayment for repayment mortgage or amount
                     saved for endowment mortgage
          RV   = rental value (imputed rent)
          OZB = property tax
          S   = non-fiscal financial recurrent subsidy

The one-off outlays (Table 10.1) and contributions (Table 10.3) could be included in the concept of expenditure in one of two ways. Either transaction costs and one-off contributions are being taken into consideration in the first year, or they are 'financed' by the mortgage loan. In the second method the one-off items are transformed into annual expenditure.

The number of components included in the outlay concept could be broadened to encompass other expenses such as maintenance. However, since the focus of this chapter is on the effects of taxation and other financial instruments on the costs of housing, these other expenses are assumed to be constant between countries, and are therefore excluded from the equation.

The expected expenditure is a liquidity concept which is indispensable to the decision 'to buy or not to buy'. The expenditure concept shows clearly how the fiscal regime affects housing outlays and the relative attraction of owner-occupation. It demonstrates the affordability of housing.

However, the expenditure concept inhibits comparisons between households because, as Hills states, it 'does not compare like with like' (1991: 46). This results from the concept's dependence on cash flows in general and on mortgage expenses in particular.

The main reason for this incomparability is the fact that the level of expenditure does not reflect the price of housing consumption. In accountancy, it is common to convert expenditure into costs, because costs are the basis for prices.

The level of expenditure however, does not take into account the opportunity costs of owner's equity. Nor are capital gains or losses included, although they directly affect the revenue attributable to a dwelling. Such gains or losses affect the owner's potential purchasing power.

Costs

In the literature concerning the economics of housing, the price of housing consumption is called 'user costs' (of (housing) capital) (e.g. Miles 1994; Megbolugbe and Linneman 1993; Hendershott 1988; Dougherty and Van Order 1982; Van Order and Villani 1982; Hall and Jorgenson 1967). Hills (1991) uses the term 'economic rent'. User costs allow comparisons to be made, between one unit of housing consumption and another. Miles refers to 'user cost . . . [as] an index of how many consumer goods needed to be given up . . . to enjoy the services of an owner-occupied home' (1992: 74). Scott states: 'The term [user cost] is currently applied to the opportunity cost of putting goods and resources to a certain use' (1953: 369). Kau and Sirmans define opportunity cost: 'In economics, the choice of one good requires the giving up of another, and the *opportunity cost* is the sacrifice of the next valued alternative' (1985: 34). Accordingly, the foregone revenue represents a cost.

The components of user cost have been identified by many economists (e.g. Conijn 1995; Miles 1994; Megbolugbe and Linneman 1993; Hills 1991; Linneman and Voith 1991; Diamond 1980; Rosen and Rosen 1980). They include:

1 *Costs of funds tied up in the house* This item comprises the costs of the owner's capital: interest on debt capital and opportunity costs on owner's equity. In classical financial theory there would be no difference between the interest rate and the opportunity cost rate (Brealey and Myers 1984).

2 *Costs of management and maintenance* These costs include insurance fees, property taxes, maintenance and transaction costs.

3 *Change in property value* The change in property value embodies three components. First, there is the (pure) price rise due to inflation. The second component concerns depreciation, the value loss due to the passage of time. Third, there are improvement investments. Though improvements generally increase house values, it is difficult to establish to what extent investments actually translate into increases in property value.

   Depreciation and maintenance can be considered as two sides of one coin (Hills 1991). While maintenance keeps up the value of a building, depreciation usually occurs at a slower rate.

Under the assumptions of no income tax, no subsidies and no improvement investments, a user cost definition expressed as percentage of house price can now be formulated (equation 2):

185

$$UC_t = P_t * (b_t*r + (1 - b_t)*i + \tau_{ozb} + m_t + d_t - \dot{p}_t) \tag{2}$$

where UC_t = user costs
        P_t = house price
        b_t = proportion of house price mortgage financed
        r = interest rate on loan
        i = interest rate on owner's equity (opportunity costs)
        τ_{ozb} = property tax rate
        m_t = managerial costs: maintenance and other costs
        d_t = depreciation
        $\dot{p}_t$ = pure price change of dwelling between t and t + 1

If, as is the case in this contribution, the effects of taxation and subsidies are the focus of attention, these effects will have to be included in the definition of user cost. The time subscripts are eliminated in equation 3 which describes the Dutch situation. The equation includes taxation of imputed rent, as well as the after-tax interest rates (of return) for equity and for debt. Return on equity is assumed completely taxable as income. If accrued capital gains/losses were relevant for income tax, they would be included in equation 3 as well. As with equation 1, maintenance is excluded here, while the taxable recurrent subsidy is included. As with the expenditure concept, the one-off contributions and costs could be included in one of two ways: either in the year in which they are incurred, or in the annual costs. The annual component could be calculated as interest on the appropriate amount of equity or debt.

$$UC = P*(b*r + (1 - b)*i + \tau_{ozb} + d - \dot{p}) - (1 - \tau_i) * S \tag{3}$$
$$+ \tau_i * P*(rv - b*r - (1 - b) * i)$$

where τ_i = income tax rate
       rv = rental value
       S = non-fiscal financial recurrent subsidy

Calculations

Calculations of expenditure and cost figures over a period of five years (1990–1994) in three of the six countries previously examined, England, Denmark and the Netherlands, now follow. As these figures are not readily available from existing research, calculations pertain to the simulation of expenditure and costs of representative samples of owner-occupiers. It is assumed that income figures and tax systems remain unchanged during the period of calculations. Simulations of price rises of the dwelling, however, are derived from the price index of residential construction (Table 10.4).

The starting point is the assumption that a household consisting of a married couple with one income purchases its home on 1 January 1990. This hypothetical household earns the average income of an owner-occupier of the country

Table 10.4 Assumed characteristics of cases (nominal amounts, 1990)

	Denmark (Dkr.)	England (£)	Netherl. (Dfl.)[a]
Gross household income (average case)	350,000	22,000	65,000
House price/income (average case)	1.60	2.70	2.90
House price	550,000	59,400	188,500
Loan-to-value ratio (%)	80	70	90
Type of loan	60% annuity, 40% linear	endowment	endowment
Loan term	30 years	25 years	30 years
Type of interest	30 years fixed (because of bond finance)	variable, but assumed constant over period	5 years fixed
Mortgage interest rate	11.01 + 0.50 costs	14.43	9.20
Yield of long-term bonds (%)[b]	10.74	11.08	8.92
Depreciation rate (assumption)	1.00	1.00	1.00
Price rise residential construction (%)[c]:			
1990	4.10	7.63	2.20
1991	4.27	3.33	2.80
1992	3.77	0.33	2.70
1993 = 1994	3.33	0.27	2.87

Source: Haffner (1998).

Notes

a 1 Dkr. = Dfl. 0.29 which makes Danish incomes much higher and prices about the same as the Dutch ones. £1 = Dfl. 3.15, which makes English incomes slightly higher and prices slightly lower than Dutch ones.

b *United Nations Monthly Bulletin of Statistics* (1995), government bonds yield is used as approximation for the before-tax return on equity.

c OECD *Historical Statistics* (1996), moving averages over three years.

reviewed. Three more cases are derived by multiplying this average by two (case 2) by 0.75 (case 3/4) and by 0.5 (case 1/2). Different income 'groups' are thus created. The income levels are less important than the relations between the cases.

The purchase price of the dwelling for each household is found by applying to household income a ratio between income and price, which is derived from national published sources. This ratio is not necessarily 'the average'. Subsequently, the purchase price is increased to include the transaction costs incurred upon buying the house to obtain the total acquisition costs. The total costs of home purchase can be combined with the financing costs which depend on the available mortgage products, the main features of which are summarised in Table 10.4.

Points of departure

The calculations using equations 1 and 3 must be adjusted according to the taxes and subsidies (outlined in Tables 10.1–10.3) applied to owner-occupation in each of the countries. As the equations show, taxation increases and tax deductions decrease expenditures and costs. For income tax, marginal household tax rates are used. This means that in fact average tax rates are being used whenever changes in income caused by deductions, for example, lead to changes in tax bracket (Hendershott and Slemrod 1983).

The property tax is assumed to be the percentage of purchase price each year as it is calculated for 1990. This includes the Community Charge in England as an approximation for the property tax which it replaced (in 1990) and as an approximation of the 'Council tax' by which it was in turn replaced (in 1993). The Community Charge comes to 2.3 per cent, 1.6 per cent, 1.2 per cent and 0.6 per cent of price for defined case 1/2 through to case 2. The decrease arises because actually there is no relationship with house value. The other two countries levy a true property tax. It comes to 1 per cent of price in Denmark and to 0.3 per cent of price in the Netherlands.

The one-off components (costs and concessions) are converted into annual outlays and costs. The fees at the point of purchase are assumed to increase the size of the loan. Denmark is the exemption. The Danish loan which is statutorily linked to the value of real estate, cannot cover transaction costs. These costs are thus to be financed by equity. The loan covers VAT however, because VAT is included in purchase price (see above). The assumption is that 80 per cent of house price relates to the building. VAT is calculated over this amount.

The non-fiscal financial instruments included in the calculations do make a difference for owner-occupiers. In contrast, the savings facility in Denmark does not make much of a difference, so the calculations ignore them. The after-tax annual contribution in the Netherlands decreases expenditures as well as costs. The same holds for Dutch loans with below-market interest rates. The one-off contribution in the Netherlands and the right-to-buy discount in England are treated differently as 'costs' and as 'expenditure' respectively. Under the outlay approach these contributions lower the amount of debt needed, whereas under the cost approach they increase equity. The annual gift is assumed to be the after-tax rate of 'return' on these contributions.

Results

Tables 10.5–10.8 present the most salient results of the calculation of housing expenditure and costs of the representative owner-occupier households. Costs and outlays are presented as percentages of price at the beginning of the year. Table 10.5 contains the results for Denmark, Table 10.6 for England and Tables 10.7 (existing dwelling) and 10.8 (newly built dwelling) for the Netherlands.

Although purchase costs in Denmark slightly differ among cases, a distinction among the cases in Table 10.5 is not necessary, as differences appear to be insignificant. Dutch purchase costs result in slightly lower percentages the higher the price of a dwelling. In England there is no difference between new and existing dwellings, as the VAT levy is 0 per cent for newly built dwellings and irrelevant for existing dwellings.

For income tax purposes, it is not necessary to distinguish between the Danish cases as they are all treated alike. This is neither the case in the Netherlands nor in England.

In general, the results for the instruments with the maximum deductions on costs and expenditure are presented in the tables. This means that the minimum right-to-buy deduction for England is excluded, as well as the Dutch one-off contribution. For the Netherlands in the case of the existing dwelling, the slight effect of the guarantee (Table 10.3) is shown. In the case of the new dwelling, this effect is excluded, but should be added to calculate the total effect of all Dutch instruments for cases 1/2 and 3/4.

Clearly, in Denmark and the Netherlands the results are influenced by type of dwelling – existing versus newly built. Expenditure as well as costs are higher for the newly built dwelling than for the existing dwelling. This stems from the fact that both dwellings are assumed to have the same price before purchase fees are added. In England this difference does not occur as VAT is not levied on dwellings.

As in all countries annual price rises have taken place (Table 10.4) and expenditure either decreases nominally (as in Denmark where annuity/linear repayment mortgages prevail) or remains unchanged nominally (as in England and the Netherlands where endowment mortgages are commonly used), expenditure decreases as a percentage of price through the years 1990 to 1994.

In the situation without assistance, pre-tax expenditure is generally highest in

Table 10.5 Nominal expenditure and user costs of owner-occupiers in Denmark (percentage of price at the beginning of the year, 1990–1994)

All cases	User costs					Expenditure				
	1990	1991	1992	1993	1994	1990	1991	1992	1993	1994
Existing dwelling:										
Total before tax	9.5	9.3	9.7	10.1	10.1	11.5	11.0	10.4	10.0	9.6
Income tax effects	−4.6	−4.6	−4.6	−4.5	−4.5	−3.4	−3.2	−2.9	−2.7	−2.5
Total after tax	4.9	4.7	5.1	5.6	5.6	8.1	7.8	7.5	7.3	7.1
New dwelling:										
Total before tax	11.5	11.2	11.6	11.9	11.8	13.3	12.7	12.1	11.6	11.1
Income tax effects	−5.6	−5.5	−5.5	−5.4	−5.4	−4.2	−3.9	−3.6	−3.4	−3.2
Total after tax	5.9	5.7	6.1	6.5	6.4	9.1	8.8	8.5	8.2	7.9

Source: Haffner (1998).

Table 10.6 Nominal expenditure and user costs of owner-occupiers in England (percentage of price at the beginning of the year), existing and new dwelling (1990–1994)

	User costs					*Expenditure*				
	1990	*1991*	*1992*	*1993*	*1994*	*1990*	*1991*	*1992*	*1993*	*1994*
Case 1/2										
Total before tax	9.3	13.4	16.3	16.4	16.4	13.8	13.0	12.6	12.6	12.6
Income tax effects	−3.4	−3.3	−3.3	−3.3	−3.3	−2.6	−2.4	−2.3	−2.3	−2.3
Total after tax	5.9	10.1	13.0	13.1	13.1	11.2	10.6	10.3	10.3	10.3
Right to buy (existing dwelling only):										
Max. deduction	−4.6	−4.3	−4.2	−4.1	−4.1	−9.0	−8.4	−8.1	−8.1	−8.0
Income tax effects	−3.4	−3.3	−3.3	−3.3	−3.3	−0.6	−0.5	−0.5	−0.5	−0.5
Total after tax	1.3	5.8	8.8	9.0	9.0	4.2	4.1	4.0	4.0	4.1
Case 3/4										
Total before tax	8.6	12.7	15.7	15.7	15.7	13.1	12.3	12.0	11.9	11.9
Income tax effects	−3.3	−3.2	−3.2	−3.2	−3.2	−2.4	−2.3	−2.2	−2.2	−2.2
Total after tax	5.3	9.5	12.5	12.5	12.5	10.7	10.0	9.8	9.7	9.7
Right to buy (existing dwelling only):										
Max. deduction	−4.6	−4.3	−4.2	−4.1	−4.1	−9.0	−8.4	−8.1	−8.1	−8.0
Income tax effects	−3.3	−3.2	−3.2	−3.2	−3.2	−0.6	−0.5	−0.5	−0.5	−0.5
Total after tax	0.7	5.2	8.3	8.4	8.4	3.5	3.4	3.4	3.3	3.4
Case average										
Total before tax	8.2	12.3	15.3	15.3	15.3	12.7	11.9	11.6	11.5	11.5
Income tax effects	−2.7	−2.7	−2.7	−2.7	−2.7	−1.8	−1.7	−1.6	−1.6	−1.6
Total after tax	5.5	9.6	12.6	12.6	12.6	10.9	10.2	10.9	9.9	9.9
Case 2										
Total before tax	7.6	11.7	14.6	14.7	14.7	12.1	11.3	10.9	10.9	10.9
Income tax effects	−2.8	−2.9	−3.0	−3.0	−3.0	−1.5	−1.4	−1.3	−1.3	−1.3
Total after tax	4.8	8.8	11.6	11.7	11.7	10.6	9.9	9.6	9.6	9.6

Source: Haffner (1998).

England, except for the case average and case 2 with a new dwelling. It is lower in Denmark and lowest in the Netherlands. This ranking is caused mainly by differing mortgage interest rates.

If prices were not to rise (a scenario not shown in the tables) pre-tax user costs in the situation without financial aid would be constant throughout the simula-

Table 10.7 Nominal expenditure and user costs of owner-occupiers in the Nether-lands (percentage of price at the beginning of the year), existing dwelling (1990–1994)

	User costs					Expenditure				
	1990	1991	1992	1993	1994	1990	1991	1992	1993	1994
Case 1/2										
Total before tax	9.2	8.5	8.6	8.4	8.4	10.3	10.0	9.8	9.5	9.3
Income tax effects	−3.3	−3.3	−3.3	−3.1	−3.1	−3.0	−2.9	−2.8	−2.8	−2.7
Total after tax	5.9	5.2	5.3	5.3	5.3	7.3	7.1	7.0	6.7	6.6
Guarantee:										
Total before tax	9.0	8.3	8.4	8.2	8.2	10.1	9.8	9.6	9.3	9.1
Income tax effects	−3.3	−3.3	−3.2	−3.1	−3.1	−2.9	−2.8	−2.8	−2.7	−2.6
Total after tax	5.7	5.0	5.2	5.1	5.1	7.2	7.0	6.8	6.6	6.5
Case 3/4										
Total before tax	9.1	8.5	8.6	8.4	8.3	10.2	10.0	9.7	9.5	9.2
Income tax effects	−3.2	−3.2	−3.2	−3.1	−3.1	−2.9	−2.8	−2.8	−2.7	−2.6
Total after tax	5.9	5.3	5.4	5.3	5.2	7.3	7.2	6.9	6.8	6.6
Guarantee:										
Total before tax	8.9	8.3	8.4	8.2	8.2	10.0	9.8	9.5	9.3	9.1
Income tax effects	−3.2	−3.2	−3.2	−3.0	−3.0	−2.8	−2.8	−2.7	−2.6	−2.6
Total after tax	5.7	5.1	5.2	5.2	5.2	7.2	7.0	6.8	6.7	6.5
Case average										
Total before tax	9.1	8.5	8.5	8.3	8.3	10.2	10.0	9.7	9.4	9.2
Income tax effects	−4.5	−4.5	−4.5	−4.4	−4.4	−3.9	−3.8	−3.7	−3.6	−3.5
Total after tax	4.6	4.0	4.0	3.9	3.9	6.3	6.2	6.0	5.8	5.7
Case 2										
Total before tax	9.0	8.4	8.5	8.3	8.2	10.1	9.9	9.6	9.4	9.1
Income tax effects	−5.4	−5.4	−5.4	−5.3	−5.3	−4.7	−4.6	−4.4	−4.3	−4.2
Total after tax	3.6	3.0	3.1	3.0	2.9	5.4	5.3	5.2	5.1	4.9

Source: Haffner (1998).

tion period. For Denmark user costs would be about 13.6 per cent for an existing dwelling (plus about 2 per cent for newly built dwellings). In the Netherlands they would be about 11.3 per cent (plus about 1.2 per cent). The costs for case 1/2 are slightly higher than for case 2, as purchase fees are proportionately higher for the former than for the latter. In England the same sort of effect is created by the Community Charge which is related to household size. User costs run from

Table 10.8 Nominal expenditure and user costs of owner-occupiers in the Nether-lands (percentage of price at the beginning of the year), new dwelling (1990–1994)

	User costs					Expenditure				
	1990	*1991*	*1992*	*1993*	*1994*	*1990*	*1991*	*1992*	*1993*	*1994*
Case 1/2										
Total before tax	10.4	9.7	9.8	9.5	9.5	11.4	11.2	10.9	10.6	10.3
Income tax effects	−3.7	−3.7	−3.7	−3.5	−3.5	−3.4	−3.3	−3.2	−3.1	−3.0
Total after tax	6.7	6.0	6.1	6.0	6.0	8.0	7.9	7.7	7.5	7.3
Recurrent contribution	−2.7	−2.6	−2.5	−2.5	−2.4	−2.6	−2.6	−2.5	−2.4	−2.4
Total	4.0	3.4	3.6	3.5	3.6	5.4	5.3	5.2	5.1	4.9
Case 3/4										
Total before tax	10.3	9.7	9.7	9.5	9.4	11.4	11.1	10.8	10.6	10.3
Income tax effects	−3.7	−3.6	−3.6	−3.5	−3.5	−3.3	−3.2	−3.1	−3.0	−3.0
Total after tax	6.6	6.1	6.1	6.0	5.9	8.1	7.9	7.7	7.6	7.3
Recurrent contribution	−1.7	−1.6	−1.6	−0.5	−0.0	−1.7	−1.6	−1.5	−0.5	−0.0
Total	4.9	4.5	4.5	5.5	5.9	6.4	6.3	6.2	7.1	7.3
Case average										
Total before tax	10.3	9.6	9.6	9.4	9.4	11.3	11.1	10.8	10.5	10.2
Income tax effects	−5.1	−5.1	−5.1	−5.0	−4.9	−4.5	−4.4	−4.2	−4.1	−4.0
Total after tax	5.2	4.5	4.5	4.4	4.5	6.8	6.7	6.6	6.4	6.2
Case 2										
Total before tax	10.2	9.5	9.6	9.4	9.3	11.2	11.0	10.7	10.5	10.2
Income tax effects	−6.1	−6.1	−6.1	−5.9	−5.9	−5.3	−5.2	−5.0	−4.9	−4.8
Total after tax	4.1	3.4	3.5	3.5	3.4	5.9	5.8	5.7	5.6	5.4

Source: Haffner (1998).

about 16.9 per cent to about 15.2 per cent for case 1/2 to case 2. The ranking from lowest to highest would be roughly: the Netherlands, Denmark and England.

Price changes greatly influence user costs. This can especially be seen in England where user costs rose steeply because of increasingly smaller price rises in the period from 1990 to 1994. In Denmark and the Netherlands the price rise for 1991 is larger than for the year before and after and this is reflected by the development of the user costs.

Including price effects, pre-tax user costs will be about 10 per cent for existing dwellings (plus about 2 per cent for newly constructed dwellings) in Denmark for 1990–1994. Price rises are relatively stable so the pre-tax user cost ranges from 9.3 per cent to 10.1 per cent. In the Netherlands, where price rises are lower than in Denmark, but are also stable, pre-tax user costs are on average 8.5 per cent (plus about 1.3 per cent) of price, excluding financial assistance. They vary from 8.3 per cent to 9.2 per cent for existing dwellings and from 9.3 per cent to 10.4 per cent for newly built dwellings. In England user costs vary greatly. Before tax, they run from 7.6 per cent to 16.4 per cent where there is no financial assistance. Generally, the ranking from lowest to highest would be the Netherlands, Denmark and England.

Income tax effects have a greater impact on costs than on outlays. The reason can be found in the components that are included. In the cost case, as well as the possible taxation of imputed value, there is also the after-tax mortgage interest and equity interest (equation 3). But equity interest is excluded from the expenditure approach.

The three countries studied present different 'fiscal' systems. In Denmark all owner-occupiers are treated equally, at least when expressed as percentage of price. The percentage of taxation of rental value and the deduction of mortgage interest is about 50 per cent for everyone. In contrast, in the Netherlands these components are deducted at the household's marginal tax rate. This results in relatively larger income tax effects for higher-income households, and in relatively lower user costs for these households. The same can be said about outlay expenses. England on the other hand, restricts the interest deduction for larger mortgages which is reflected in expenditure and costs. This reasoning is of course based on the assumption that 'richer' households have larger mortgages than 'poorer' households.

Household tax rate in Denmark is comparable to the tax rate of the Dutch case average and case 2. These Dutch and Danish cases are the cases with the relatively largest income tax deductions, as mortgage tax relief is not limited in amount.

A ranking of countries for total costs and expenditure cannot be made. Not only do fiscal effects have a role, depending on income level, but price effects are important also. If there were no price rises, the ranking from lowest to highest would be as before: the Netherlands, Denmark and England. What one will be able to observe, if price rises are included is that especially the Dutch case 2, but also the Dutch case average, have the lowest user costs of all countries throughout the years. This is due to the income tax treatment.

From the non-fiscal concessions that reduce housing expenditure and costs (Table 10.3), the right-to-buy reductions are very generous. The discount results in a reduction of costs anywhere between 2.5 per cent (the minimum discount, not shown in tables) and 4.6 per cent of price in the first year. Expenditure can even be reduced up to 9 per cent. The Dutch recurrent contribution reduces costs by 2.7 per cent in the first year and expenditure by 2.6 per cent for case 1/2. Expenditure and costs for the Dutch case 3/4 increase in the fourth year as the recurrent contribution decreases and ends altogether in the fifth year (Table 10.8). The loan with a below-market-interest rate of 0.2 per cent (not shown for newly built dwellings) reduces costs and expenditure by 0.1 per cent to 0.2 per cent.

Conclusion

The application of the expenditure and user cost concepts to owner-occupiers has shown both to be useful for different purposes. The expenditure concept is a liquidity concept which is essential for the decision 'can we afford to buy or not to buy?'. This concept clearly shows how the interplay between mortgage products and the fiscal regime in a country hampers or facilitates home ownership. However, direct comparison of housing outlays among countries is not feasible using this measure. The crux of the comparability problem lies in the treatment of the mortgage expenditure. This determines the level of expenditure without being an expression for the price of housing consumption.

In contrast, the user cost of capital concept must be applied for an equivalent cost comparison between owner-occupiers. User costs present the price of housing services or consumption. The most important distinction from the expenditure concept is the inclusion of the costs of the capital invested in the house and of the change of house value, and the exclusion of the mortgage repayments.

The simulation calculations for Denmark, England and the Netherlands compare costs and expenditure across countries for 1990–1994. Comparisons are based on totals as percentage of price. Calculations show that user costs are generally lower than expenditure, unless house prices rise very little or even fall for the first few years after the purchase. Income tax effects on the other hand are higher on a cost basis than on an expenditure basis, due to the definitions used.

The different income tax systems are reflected in the costs and expenditure: equal treatment in Denmark, relatively favourable treatment of low income households in England and comparably favourable treatment of the high income households in the Netherlands.

The ranking of countries for total costs and expenditure cannot be achieved without complications. Not only do fiscal effects have a role, depending on income level, but price effects are important too. The Dutch case 2, the highest income case (but also the Dutch case average), however, has (have) the lowest user costs percentage of all countries during the period reviewed.

Of the various non-fiscal concessions offered by governments to owner-occupiers, it has been shown that the English right to buy is very effective in lowering costs and cash flows for tenants buying their council homes. The effect of the Dutch recurrent contribution (for newly built dwellings) is much less than under the right to buy. Given that the Dutch scheme has now been abolished, the generosity of the right-to-buy scheme becomes still clearer.

Although the combined impact of fiscal and non-fiscal concessions on costs and expenditure can be estimated, nothing can be said about the amount of housing subsidies per household. According to Ermisch, a subsidy is 'a divergence between the cost of housing faced by the consumer and the cost of producing that housing' (1984: 3). To measure such a divergence, a subsidy benchmark is necessary: how much should the costs of housing consumption be? The above calculations only present actual housing costs/expenditure.

Often subsidisation estimates are based on the effects of income taxation on housing expenditure. As the calculations demonstrate, the price of housing consumption results in a 'fiscal' effect which differs from the fiscal effect based on the expenditure concept. If the income tax effects are presented as the only type of fiscal 'subsidisation', the economic cost approach leads to a higher estimate of subsidy than the expenditure approach. However, it should be noted that concessions in purchase fees, notably VAT, represent a subsidy which is not reflected in this way.

To make statements about subsidy, the term must first be defined. Even though the definition might be arbitrary, it may make possible a ranking of households according to the amount of subsidy received.

Implications for European integration

From the discussion, one can infer that the (European) integration of the definitions of housing costs and subsidies is still far from being accomplished. How costs for a presumed identical unit of housing services 'should' be estimated, is open to debate. A common definition and measurement of housing subsidies is a goal which is also very far from being attained. On housing expenditure, slightly more information than on costs and subsidies is available at a European level, especially through national housing and budget surveys. However, comparison across households about the level of expenditure may not always be useful.

Integration of these matters calls for agreement about definitions, after which the collection of comparable data becomes possible. This chapter is intended to be a contribution in clarifying the concepts of housing expenditure and cost.

References

Brealey, R. and Myers, S. (1984) *Principles of Corporate Finance*. Auckland: McGraw-Hill.

Conijn, J.B.S. (1995) *Enkele financieel-economische grondslagen van de volkshuisvesting* (Some Financial-Economical Principles of Housing), reeks Volkshuisvestingsbeleid en Bouwmarkt, 25. Delft: Delft University Press.

De Kam, F. (1997) 'De logica van de hypotheekrente-aftrek', *Economisch Statistische Berichten*. 82, 4091: 94–95.

Department of the Environment, (1990) *Housing and Construction Statistics 1979–1989*. London: HMSO.

Diamond, D.B. (1980) 'Taxes, Inflation, Speculation and the Cost of Homeownership', *AUREUEA (American Real Estate and Urban Economics Association) Journal* 8: 281–298.

Dougherty, A. and Van Order, R. (1982) 'Inflation, Housing Costs and the Consumer Price Index', *American Economic Review*. 72, 1: 154–164.

Ermisch, J. (1984) *Housing Finance: Who Gains?* London: Policy Studies Institute, Family Income Support Part 3, no. 628, Studies of the Social Security System no. 5, June.

Haffner, M.E.A. (1992a) *Fiscus en eigen-woningbezit in de EG, deel 2: Berekening van de woonuitgaven* (Taxation and Owner-Occupied Housing in the EC, Part 2: Calculation of Housing Expenses), internal report. Delft: OTB Research Institute for Policy Sciences and Technology.

Haffner, M.E.A. (1992b) *Eigen woning in de EG: Fiscale en overige financiële instrumenten* (Owner-Occupied Housing in the EC: Fiscal and Other Financial Instruments), reeks Volkshuisvestingsbeleid en Bouwmarkt 18. Delft: Delft University Press.

Haffner, M.E.A. (1993) 'Fiscal Treatment of Owner-Occupiers in Six EC Countries: A Description', *Scandinavian Housing and Planning Research* 10, 1: 49–54.

Haffner, M.E.A. (1994) 'Effects of Financial Policy on Housing Expenses of Owner-Occupiers of Newly Built Dwellings: A Comparison of Six Countries in North-western Europe', *Housing Studies* 9, 1: 125–141.

Haffner, M.E.A. (1998) dissertation, forthcoming, Delft: Delft University Press.

Hall, R.E., and Jorgenson, D.W. (1967) 'Tax Policy and Investment Behavior', *American Economic Review*, 57: 391–414.

Hendershott, P.H. (1988) 'Home Ownership and Real House Prices: Sources of Change, 1965–1985', *Housing Finance Review*, 7: 1–18.

Hendershott, P.H., and Slemrod, J. (1983) 'Taxes and the User Cost of Capital for Owner-Occupied Housing', *AREUEA (American Real Estate and Urban Economics Association) Journal* 10, 375–393.

Hills, J. (1991) *Unravelling Housing Finance: Subsidies, Benefits, and Taxation.* Oxford: Clarendon Press.

Kau, J.B. and Sirmans, C.F. (1985) *Real Estate.* London: McGraw-Hill.

Linneman, P. and Voith, R. (1991) 'Housing Price Functions and Ownership Capitalization Rates', *Journal of Urban Economics* 30, 1: 100–111.

Megbolugbe, I.F. and Linneman, P.D. (1993) 'Home Ownership', *Urban Studies* 30, 4/5: 659–682.

Miles, D. (1992) 'Housing and the Wider Economy in the Short and Long Run', *National Institute Economic Review* Feb.: 64–78.

Miles, D. (1994) *Housing, Financial Markets and the Wider Economy.* New York: John Wiley and Sons, Wiley Series in Financial Economics and Quantitative Analysis.

Ministerie van VROM (1989) *Volkshuisvesting in de jaren negentig: van bouwen naar wonen* (Housing in the 1990s), Tweede Kamer (Parliament), vergaderjaar 1988–1989, 20 691, nrs. 2–3, 's-Gravenhage: SDU uitgeverij.

OECD (1996) *Historical Statistics.* Paris: OECD.

Rosen, H.S. and Rosen, K.T. (1980) 'Federal Taxes and Homeownership: Evidence from Time Series', *Journal of Political Economy* 11: 1–23.

Scott, A.D. (1953) 'Notes on User Cost', *Economic Journal* 63 (June): 368–384.

Taffin, C. (1991) 'Accession: L'ancien réhabilité', *Economie et Statistique* 240: 15–17.

UN (1995) *Monthly Bulletin of Statistics* 49, 12. New York: UN.

Van Order, R. and Villani, K. (1982) 'Alternative Measures of Housing Costs', in C.F. Sirmans (ed.) *Research in Real Estate: A Research Annual,* Vol. I. Greenwich, CT: Jai Press Inc.

11

HOMELESSNESS IN THE EUROPEAN UNION

Suzanne Fitzpatrick

In this chapter, I provide an overview of homelessness in the EU in the context of the EU's concern with social cohesion and exclusion. I look at: the definition and nature of homelessness; the extent of homelessness; explanations of homelessness; the characteristics of homeless people and trends in the homeless population; and rights and services for the homeless.

The main theme of the chapter is to identify similarities and distinctions in the national homelessness situations across the EU. There is a particular focus on the quite different ways in which homelessness is conceptualised and explained in different countries, and whether these distinctions reflect genuine differences in the phenomenon or simply different political and scientific perspectives. The chapter ends with a discussion of the implications of this analysis for the role of the EU in tackling homelessness in the member states.[1]

Homelessness and social exclusion

'Social cohesion' became an objective of the European Community by virtue of the Single European Act 1987 (Article 130 A-E). This has since been interpreted as a competence to combat 'social exclusion' (House of Lords 1994). The European Commission argues that social exclusion is a more dynamic concept than poverty, referring to processes as well as outcomes, and is also broader, relating to marginalisation in all aspects of social, political and economic life, as well as income deprivation. The House of Lords Select Committee on the European Communities provides a useful summary:

> income poverty and social exclusion are not co-terminous: poor housing and health care, unemployment and lack of training, old age and single parenthood may all interact and react with low income to restrict economic and social opportunities with the result described as social exclusion.
>
> (House of Lords 1994: 7)

Homelessness is perhaps the most extreme manifestation of social exclusion, representing the denial of a fundamental requirement of social integration: adequate shelter. Therefore as a marginalised group the homeless are entitled to help under European Union programmes to combat social exclusion. However to qualify for funding from these Structural Funds a homelessness project must combine education, training or work creation with accommodation, and must not be concerned solely with the rehousing of homeless people. EU funding has been granted for homelessness projects under the Horizon programme which provides training and job creation for disadvantaged groups. Funds have also been secured for the foyer hostel network which offers accommodation and training to young people (Stephens *et al.* 1997). However such EU assistance to homeless people has thus far been very limited in scale.

Definition and forms of homelessness in the EU

There is a wide range of official and unofficial definitions of homelessness employed within the EU, reflecting different national perspectives on the issue, and the views of various commentators whose interests are served by either minimising or maximising the scale of the problem. In the absence of any agreed common definition of homelessness, the European Federation of National Organisations Working with the Homeless (FEANTSA) offers the following working definition:

1   People who are roofless.
2   People who are living in institutions because they have no other place to go.
3   Those living in insecure accommodation.
4   Those living in entirely substandard or inappropriate accommodation (Harvey 1994).

This definition offers a useful broad outline but is clearly open to a great deal of subjective interpretation. For instance, how 'substandard' must accommodation be to constitute homelessness? One must always be careful in attempting to include all groups who may legitimately be regarded as homeless, so that the term does not become over-extended to the point where it includes everyone who is not adequately housed, and therefore entirely useless. The above typology also fails to capture the dynamics of homelessness: it does not distinguish, for instance, between the completely dissimilar situations of the short-term and the chronically homeless. Furthermore, it is clear that the concept of homelessness, like 'home' from which it is derived, is not a purely housing based concept and has social and psychological dimensions which are not acknowledged in FEANTSA's definition. However, it must be conceded that any definition which attempted to incorporate all these meanings of homelessness would become too complex and abstract to be of any practical use, and it is perhaps inevitable that a working definition be reduced to the housing dimension.

One further point should be made regarding the definition of homelessness. FEANTSA, like many other commentators, has argued that homelessness is a relative rather than absolute term, to be judged by the standards prevailing in a particular society at a particular time (Daly 1992). This position seems sensible, and sits well with the concept of social exclusion. However it would appear to undermine FEANTSA's call to the national governments of the EU to adopt a common definition of homelessness given the still widely varying housing and social conditions found across Europe.

We now turn to the specific types of accommodation homeless people occupy in the EU, and the similarities and differences between countries in the form which homelessness takes.

Rooflessness is apparent in major cities throughout the EU, but there are some important differences between northern and southern Europe. Significant numbers of 'street children' and roofless families can be found in Spain, Portugal, Greece and in some parts of Italy. In the northern countries most roofless people are single adult males, although single women and young people form an increasing proportion of rough sleepers. Roofless children or families are exceptional in northern Europe (Drake 1994).

There are many types of homeless accommodation common to most countries in the EU. Hostels, bed and breakfast hotels and guest houses provide shelter to homeless people throughout Europe. In all member states there are homeless people squatting in disused buildings and living in caravans and tents because they have no proper accommodation. Similarly, in all countries within the EU there are people staying in institutions such as psychiatric hospitals and prisons who should be considered homeless because they have nowhere to go on their release. Indeed, many of these people will have become institutionalised because of their homeless situation.

In all member states of the EU there are people living in severely overcrowded accommodation, or sharing housing with friends and relatives on a permanent or temporary basis because they have no home of their own. However it may be that cultural and social conventions differ from country to country within Europe so that conditions which are commonplace and socially accepted in one member state, may be viewed as extreme deprivation and therefore homelessness in another.

Shanty-towns are now found only in southern Europe. These settlements of rudimentary dwellings constructed by the dwellers themselves out of materials such as cardboard or corrugated iron are evident in urban areas of Spain, Portugal and Greece. For example, the 1991 Census in Portugal recorded 16,104 'barracus' (shanty-town huts) heavily concentrated around the metropolitan areas of Lisbon and Oporto (Neves 1995). FEANTSA has estimated that around 60,000 people live in Portuguese shanty-towns (Drake 1994). The Portuguese government is now engaged in a programme to construct 15,000 houses to rehouse the shanty-town dwellers, and aims to eliminate all 'barracus' in Lisbon and Oporto by 2001 (Neves 1995). Shanty-towns have also reappeared in Spain, and are occupied mainly by gypsies and immigrants from North Africa (McCrone and Stephens 1995).

There are shanty-towns of 'precarious houses' on the outskirts of cities throughout Greece, although they are now less prevalent than a few decades ago (Sapounakis 1994). These dwellings are built either by persons squatting on land owned by the Greek state or, more typically, by persons who have bought land but are unable to build a proper dwelling. This may be because they cannot obtain a legal permit, or because their housing needs are too urgent to await the construction of a conventional house. It should be noted that shanty-towns are a common method for starting a new settlement in Greece, and in other countries such as Turkey, and often develop eventually into adequate housing.

In contrast to these shanty-town dwellings in Mediterranean Europe, even the most substandard housing in a northern country such as the UK or Germany will be solidly built, although it may be in such poor condition that persons living there are considered homeless. The far harsher climate in northern Europe would, of course, make it very much more difficult to survive in shanty-town conditions.

The extent of homelessness in the EU

There is little doubt that homelessness spread alarmingly in the EU member states from the mid-1980s onwards, but there is an absence of reliable figures on the overall numbers of homeless people. FEANTSA provides estimates of the extent of homelessness based on the reports of its national correspondents. These suggest that by the early 1990s:

1   on an average day, around 1.1 million people in the EU had to sleep rough, squat or had to rely on public and voluntary services to provide them with accommodation;
2   over the course of a year at least 1.8 million homeless people in the EU were dependent on public or voluntary services for temporary shelter, and an additional 0.9 million rotated between staying with friends and relatives, living in short-term private accommodation and using services for homeless people;
3   a further 15 million badly housed people were living in severely substandard and overcrowded dwellings (Avramov 1995).

The most recent national figures are presented in Table 11.1 and suggest that EU member states can be clustered into three main groups as regards their incidence of homelessness. The highest rates of homelessness, of around 10 per 1,000 inhabitants, are found in the three largest countries: France, Germany and the UK. There is a second group of member states, including Belgium, Italy and the Netherlands, which have lower rates of homelessness of around 2 per 1,000.

Table 11.1 Estimated number of people dependent on services for the homeless in member states in the early 1990s

Country	On an average day, or on the day of a survey	Over the course of a year
Austria	6,100	8,400
Belgium	4,000	5,500
Germany	490,700	876,450
Denmark	2,947	4,000
Spain	8,000	11,000
Finland	4,000	5,500
France	250,000	346,000
Greece	5,500	7,700
Ireland	2,667	3,700
Italy	56,000	78,000
Luxembourg	194	200
Netherlands	7,000	12,000
Portugal	3,000	4,000
Sweden	9,903	14,000
UK	283,000	460,000
Total	1,133,011	1,836,450

Source: Avramov (1996: 77).

At the lowest end of the spectrum are Denmark, Portugal, Spain and Greece (Harvey 1994).

However, such data is of poor quality because of the difficulties inherent in measuring the extent of homelessness in the EU. First, the absence of a commonly agreed definition means that the basis for measurement is different in each country. Second, technical difficulties exist with the available data. For example, some national statistics relate to the number of homeless people at a particular point in time, while others relate to the flow of people becoming homeless over a period of time. Also, figures given by some countries relate to homeless households, but others to homeless people. Third, the short-lived and hidden nature of many people's homelessness makes it a difficult phenomenon to quantify. Therefore most estimates of homelessness only include people who have been recorded as having sought help from agencies working with the homeless.

Avramov (1995) attempts to overcome some of these difficulties by adjusting FEANTSA's country-estimates according to a number of research-based hypotheses in order to provide minimally comparable figures for member states. It is these adjusted estimates which I have quoted in this chapter. Whilst the adjusted figures probably do represent an improvement upon the crude national statistics, they remain, as Avramov acknowledges, very rough estimates, as the hypotheses are somewhat dubious and the baseline data is unreliable. These statistics can, however, be used with reasonable confidence as minimum figures, because, on balance, the flaws in the data tend to underestimate the incidence of homelessness. But there are severe difficulties with transnational comparisons as I shall now explain.

It would appear from FEANTSA's figures that homelessness is most prevalent in the northern and in the larger countries, and least prevalent in the southern and in the smaller countries. However this is likely to be misleading. Whilst effective preventative policies in some small northern countries, such as Denmark, may genuinely limit the extent of homelessness, low numbers may also reflect poor reporting mechanisms and narrow definitions of homelessness (Avramov 1995). As estimates are based mainly on numbers of service users, these statistics will often represent the relative level of service provision in each country rather than the level of need. Therefore, paradoxically, member states with the best developed services for homeless people will record the highest levels of homelessness, and the low proportions of homeless people reported in the southern countries may be due primarily to the lack of service provision there.

However it may be the case that homelessness is a newer and less prevalent phenomenon in southern than in northern Europe, perhaps because of the distinct social and cultural context. This is certainly the view of FEANTSA's Greek correspondent:

> In Greece as indeed in most Mediterranean countries the development of homelessness has had a noticeable delay. The phenomenon of homeless people wandering around the big cities is fairly new. The general public is not familiar with such scenes and tends to believe that the phenomenon of homelessness does not exist in Greece or at least that it is not typical of the Greek lifestyle.
>
> (Sapounakis 1994: 23)

On the other hand, in southern countries with few public services for homeless people there is likely to be greater reliance on 'private' coping strategies such as staying with friends and relatives, or staying in unconventional accommodation such as tents and caravans. Shanty-towns may absorb homelessness which takes other forms in northern Europe. Avramov (1995) has argued that if these groups, who in most countries would be considered homeless, were included in the estimates for the southern member states the numbers would rise to at least 160,000 in Spain, to 67,000 in Greece and to almost 100,000 in Portugal. It must also be remembered that, unlike northern countries, the roofless population in these southern member states includes a significant proportion of children and families.

Harvey (1994) has observed that the number of people recorded as homeless in Europe has continued to rise in the 1990s. He reported that the annual growth rate of homelessness in 1993 was 8 per cent in the Netherlands, 16 per cent in Luxembourg and 20 per cent in the UK, and argued that if we were to extrapolate these trends the number of homeless people in the twelve EU member states would be 6.6 million by 2001. However homelessness trends will adjust according to the prevailing social, economic and political climate in member states, and so such projections are unlikely to be accurate. Indeed, Maclennan *et al.* (1997) report that policy action and/or an easing of demographic pressures are now leading to reductions in homelessness in the UK, Ireland and Denmark.

To summarise, the true extent of homelessness in the EU is unknown, but the statistics provided by FEANTSA offer a minimum figure for the homeless population which demonstrates the serious and widespread nature of the problem. Whilst the available evidence suggests that homelessness is continuing to grow in many parts of the EU, there does appear to be a levelling off in some countries although the totals remain worryingly high. Transnational comparisons of rates of homelessness cannot be made with any confidence on the basis of the available statistics, and would require standardised reporting mechanisms. It would probably be very difficult, if not impossible, to introduce such mechanisms, particularly given the widely differing forms which homelessness takes in the various member states.

Explaining homelessness in the EU

There is a range of social and economic trends evident throughout the EU which have been associated with growing social exclusion and rising levels of homelessness. The main factors are outlined below.

Changing housing markets

Increasing pressure on the supply of affordable, rented accommodation for people on low incomes is a common theme throughout much of the EU. A decrease in investment in social rented housing in northern European countries has made it more difficult for new households to gain access to cheap accommodation. Expansion of the owner-occupied sector and deregulation of the private rented sector do not compensate for the diminishing stock of public housing because this private accommodation is often too expensive for those on low incomes. The growth of the private rented sector evident in a number of European countries seems to offer an alternative tenure to those who would otherwise be owner-occupiers, rather than to people who would have entered the social housing sector (Maclennan *et al.* 1997).

At the same time, demographic and social trends have led to a rise in the number of households requiring accommodation. There was an increase of 17 per cent in the number of households in the EU between 1981 and 1991, in comparison to a rise of only 2.8 per cent in the population (Harvey 1994). The housing stock expanded by only 9.5 per cent over this period. Therefore this increase in demand for housing, particularly single-person accommodation, has placed further pressure on the housing system, producing acute tensions at the bottom end of the market.

Poverty and unemployment

Homelessness is strongly associated with poverty and unemployment. Unemployment climbed to over 10 per cent of the European workforce in the mid-1990s, and the rate of youth unemployment rose even more alarmingly to over 20 per cent (Eurostat Yearbook 1995). At 45 per cent Spain had the highest recorded rate of youth unemployment (Eurostat Yearbook 1995). Changes in the structure of the labour market have meant that those who are employed often have temporary, insecure or part-time work.

There has also been a retrenchment of welfare states across much of the EU, although there are a few notable exceptions such as Spain where limited welfare provisions have been expanded slightly. Portugal and Greece lack income support provisions and Italy and Spain have very underdeveloped systems. Even elsewhere in the EU where there are relatively well-developed welfare states, sections of the population may be excluded from income support, or the rates paid are so low that recipients cannot sustain accommodation (Drake 1994).

Family fragmentation and individualisation of society

Family breakdown and social isolation are important aspects of homelessness in all member states, and the growth in the homeless population has been associated with a range of trends affecting family structures. The marriage rate in the fifteen EU member states has fallen from 7.9 per 1,000 in 1960 to 5.2 per 1,000 in 1994 (Eurostat Yearbook 1996). At the same time the divorce rate has spiralled upwards, with the total number of divorces in the EU virtually quadrupling since 1960 to reach almost 650,000 by 1993 (Eurostat Yearbook 1996).

Partly as a result of these trends, there has been a substantial growth in the number of single-person households in all EU member states, but this has been much more pronounced in the northern countries. In Germany, for example, the proportion of households containing a single person rose from 12 per cent in 1950 to 34 per cent in 1991. There has also been a marked increase in single-parent families. For example, the proportion of families with children headed by a lone parent grew in Belgium from 9.4 per cent in 1981/82 to 14.6 per cent in 1990/91, and in the UK from 13.7 per cent to 19 per cent. Lone parents are less common in southern Europe, but the trend here is also an upward one.

In the southern member states, the more recent 'individualisation' of society has been held responsible for the weakening of familial and social networks which traditionally provided the main source of social welfare. The consequences have been particularly severe in countries such as Italy which have witnessed 'increasing fragility of social relationships' on the one hand, and the 'dismantling of the welfare state' on the other (Tosi 1995: 20).

Migration patterns

Levels of migration into the EU have risen in the 1990s, with over 1 million people entering the EU each year in the early 1990s (Eurostat Statistics in Focus 1995). This increase in immigration has been closely associated with political instability in neighbouring states, particularly the break-up of the Communist bloc and conflict in the former Yugoslavia, and is unlikely to be sustained in the longer term. Germany, Italy and Greece have been the countries most affected (Harvey 1994).

De-institutionalisation

Policies to de-institutionalise psychiatric units have been pursued in many northern member states, often without adequate provision of community care facilities for those who have been discharged. The result has been an increase in the number of people with mental health problems becoming homeless.

How do these trends translate into homelessness? As Drake (1994: 2) commented: 'It is difficult to prove a direct causal link between processes occurring within society and the homeless population.' One point made by a number of commentators is that a combination of economic and social forces produces homelessness, and the phenomenon is not reducible to one simple explanatory factor. As the Italian correspondent says 'Rather than any specific single factor, growing vulnerability [to homelessness] turns out to be an outcome of the combination of various processes – unemployment *and* housing difficulties *and* demographic changes' (Tosi 1995: 10, original emphasis). A range of structural processes operating in different spheres push people to the fringes of society where they become vulnerable to social exclusion and homelessness. This would explain why homelessness expands both in areas of plentiful housing and high levels of unemployment, such as Glasgow, and in areas of pressurised housing markets but plentiful employment, such as London in the late 1980s. Therefore whilst it is clear that the housing market is an important factor in the generation of homelessness, it is equally clear that homelessness is not 'simply a housing problem' (O'Sullivan 1994: 11).

The explanation of homelessness is even more complex than the combination of structural forces discussed above. Another set of issues comes into play: individual factors. These include personal problems and circumstances which make individuals particularly susceptible to homelessness such as alcohol or drug abuse, social isolation, poor family relationships, bad health, mental illness, and histories of sexual or physical abuse. Of course, the experience of homelessness will often exacerbate or create such personal problems.

There is no clear-cut division between structural and individual factors and many issues could be interpreted as operating at either level: for instance, social isolation could be viewed either as a personal problem or as the result of growing individualisation in society. Also, there is a great deal of interaction between individual and structural factors and causal ambiguity between the two levels. For example, unemployment may lead to alcoholism and also vice versa. However, the notion of structural and individual factors is a useful broad distinction which aids our consideration of the nature of homelessness.

Virtually all commentators acknowledge that both structural and individual factors are relevant to the production of homelessness. But it is notable that various FEANTSA national correspondents placed a radically different emphasis on one or the other as the key to explaining homelessness in their countries.

The German correspondent conceptualised homelessness almost entirely in relation to the shortcomings of the housing market and the way in which it excluded groups of people from 'normal housing' (Specht-Kittler 1994). In fact he explicitly redefined homelessness as 'houselessness', and made almost no mention of social or personal problems among the homeless population. Similarly, the Irish correspondent laid heavy emphasis on homelessness as the outcome of macro-processes in Irish society, in particular its inegalitarian nature. For example, he explained youth homelessness as 'a manifestation of structural inequalities in Irish society [which] . . . can only be alleviated by redressing these structural inequities' (O'Sullivan 1994: 6).

The approach of the Danish and Dutch correspondents contrasts sharply with the above explanations of homelessness. The Danish correspondent explicitly rejects the notion that housing supply or access is the key to solving homelessness in his country because 'for the majority of Danish homeless the abode does not seem to be the major problem' (Rostgaard et al. 1995: 48). Whilst conceding that some homeless people in Denmark could manage on their own if they were able to find a cheap place to stay, he argues that research has shown that the majority of homeless people in Denmark have alcohol abuse or mental illness as their 'dominant problems'.

The Dutch correspondent offered a seven-fold typology of homeless people which was based almost entirely on personal issues such as drug addiction, psychiatric illness and anti-social behaviour. He refers with approval to another Dutch researcher's analysis that:

> a homeless person has a number of biographical characteristics namely: a problematic childhood, affective and pedagogical neglect, disturbed parental ambience, relation[ship] problems, alcohol and drug abuse and involvement in crime.
>
> (Feijter and Radstaak 1994: 11)

Again, there is an explicit emphasis on homelessness stemming from acute personal problems rather than the housing market or other structural factors.

Resolving these tensions between predominantly structural and predominantly individual explanations of homelessness is crucial to our understanding of the phenomenon in the EU. This debate has important implications both at a conceptual and at a practical level. The Italian correspondent was particularly anxious to clarify theoretical matters and was concerned that the focus of FEANTSA's 1994 national reports risked inducing a 'reductive/uncontrolled identification of homelessness in terms of housing problems' (Tosi 1995: 55). The way in which we conceptualise homelessness determines how we as a society view homeless people, and what policies and services we believe to be appropriate in combating the problem. Put crudely, if homelessness is primarily a housing problem then we must supply more housing to those in need, but if homelessness stems mainly from personal problems then more social support should be the priority. The key issue for this chapter is whether the different positions on this matter taken by FEANTSA national correspondents simply reflect different political positions and/or scientific approaches, or are the result of genuine differences in the nature of the phenomenon in different countries.

It seems likely that the most vulnerable members of society are the first to suffer in the face of negative structural forces, such as a tightening housing market. Therefore persons with the sort of personal problems highlighted by the Dutch and Danish correspondents will be at greatest risk of homelessness. A fuller discussion of the characteristics associated with homelessness is provided below. However, the more pressure there is on a housing market, the wider the cross-section of the population who will be squeezed out, including many who simply need accommodation and have no additional personal problems. On the other hand, no matter how effectively a housing market operates, there will probably always be a proportion of the population who cannot sustain housing for personal reasons.

Therefore a reasonable hypothesis would be that in a well-housed society with generous social security policies, such as Denmark, the overall levels of homelessness will be low, and a high proportion of those who find themselves homeless will have complex personal problems (Avramov 1995). Whereas a country like Germany with acute pressures on its housing market will have much higher levels of homelessness, and a lower proportion of homeless people with additional personal problems. Unfortunately the available comparative data on EU countries does not allow us to test this hypothesis properly, but the above discussion does suggest that homelessness is likely to be a significantly different phenomenon in the various member states of the EU.

Homeless people's characteristics

There are key characteristics of homeless people and trends in the homeless population evident throughout the EU. The most important of these are discussed below.

Family situation

Solitude is an important characteristic of homeless people. They are much more likely than the general population never to have married or to be divorced or separated. The Observatory estimates that 90 per cent of homeless men and 60 per cent of homeless women have no current partner (Harvey 1994). Single people are estimated to have a four times greater risk of being homeless than couples with or without children, and lone parents have a seventeen times greater risk (Daly 1993). Homeless people are often isolated and lack the sort of social and familial support which other people can rely on to help them deal with crises in their lives.

However, Drake (1994) considers that the Observatory lacks information on family homelessness because most of those surveyed for the national reports were living in hostels catering mainly for single people. She therefore argues that the above results regarding family situation represent only a partial picture of the European homeless population.

Age

The average age of the homeless appears to be falling. More than half the people in shelters for the homeless are between 20 and 39 years old, yet this age group accounts for less than one-third of the total population of the EU (Avramov 1995). There has been a dramatic escalation in youth homelessness in recent years as the transition into adulthood and independent living has become much more hazardous for many young people. Street children are becoming increasingly visible in Spain and Portugal.

Homeless people tend to be younger in the southern member states than in the north, and homeless women are, on average, younger than homeless men. The numbers of elderly homeless are very small, but this may reflect high death rates among homeless people (Drake 1994).

Gender

The available figures suggest that most homeless people are men. Women account for between a fifth and a third of the homeless in most countries. However, the proportion of women in the homeless population does appear to be rising throughout the EU. In Denmark, for example, women accounted for only 6 per cent of homeless people seeking shelter in 1976 but 20 per cent in 1989 (Harvey 1994).

Many commentators attribute the gender imbalance in these figures at least partially to women's homelessness being more 'concealed' and thus less easily enumerated than men's. Thus Daly (1993: 7) argues that women are '*more likely* than men to seek a "private" solution to their homelessness, by for example getting temporary accommodation from a friend or family member' (my emphasis). Whilst evidence does exist to suggest that there are a great many women in such 'hidden' homeless situations (Webb 1994), this is not equivalent to establishing that there are more hidden homeless women than men. I know of no research which compares hidden homelessness amongst men and women and therefore could substantiate the claim that men are *less* likely than women to seek a 'private' solution to their homelessness.

Homelessness amongst women is closely related to personal relationship problems, and a great many homeless women have suffered physical or sexual abuse (Harvey 1994). A high proportion of homeless women in all countries are accompanied by children, and practically all homeless lone-parent families are headed by a woman (Daly 1993).

Employment and income

The homeless come overwhelmingly from backgrounds of poverty. Only very small proportions of homeless people are employed, ranging from 5 per cent in Germany to 15 per cent in Belgium (Harvey 1994). Those who are unemployed tend to have casual unskilled work and low pay. However the proportion of homeless people who have never worked is very small (Harvey 1994).

With poor work records, and therefore insufficient social insurance contributions, the main source of income for homeless people in the EU is state income support. Nevertheless it is alarming how many homeless people are not in receipt of benefit, even in northern European countries. In Belgium, for example, 20 per cent of shelter residents did not receive any social assistance. Luxembourg has by far the most serious situation amongst the northern states with only 23 per cent of homeless people receiving income support. The southern European states generally have less developed welfare states and correspondingly low proportions of homeless people receiving state benefits. For example, in Italy only 29 per cent of homeless people received social assistance or a pension, 13 per cent were receiving assistance from a private agency and 43 per cent were dependent on begging and theft (Drake 1994).

Health

Homeless people have poorer health than the housed population. One-third of homeless people surveyed in Spain had serious health problems and 39 per cent of homeless people in Italy had some form of health problem (Drake 1994). There are higher than normal levels of alcohol and drug addiction found among the homeless population, for instance, a census carried out in Denmark in 1988 found that 56 per cent of homeless people surveyed had a substance abuse problem (Drake 1994), although, for the reasons discussed, we would expect an exceptionally high proportion of homeless people in countries such as Denmark to have additional personal problems. Avramov (1995) argues that proportions of homeless people with personal problems such as mental illness or substance abuse are often overestimated because research focuses on rough sleepers and hostel dwellers rather than the broader homeless population. FEANTSA also contends that for the majority of homeless people substance abuse followed rather than preceded homelessness (Harvey 1994).

Mortality rates provide the starkest evidence of poor health amongst the homeless. In Germany homeless people have a ten-year lower life expectancy than the rest of the population, and research on rough sleepers in London found that the average age of deceased homeless persons was only 47 as compared with the normal life expectancy of 73 years for men and 79 years for women (Daly 1993).

It is clear that a homeless lifestyle, particularly sleeping rough, carries great risks to health. However there is also some evidence that ill-health itself, because of its links to poverty, may also be a factor precipitating homelessness (Harvey 1994).

Experience of institutions

There is a clear relationship between homelessness and experience of institutions such as prison, psychiatric hospitals and children's homes: 75 per cent of homeless people in Ireland, 60 per cent of Belgium's homeless and 42 per cent of Italy's homeless had stayed in an institution at some stage in their lives (Drake 1994). The link between de-institutionalisation policies and increasing numbers of homeless people with mental health problems has already been mentioned. Evidence from England and Wales suggests that between 30 and 50 per cent of those sleeping rough there have a background of mental illness (Harvey 1994).

Immigrants

Whilst the vast majority of the homeless are nationals of the country in which they are homeless, immigrants do run a higher risk of homelessness than the rest of the European population. It is estimated that between 10 and 20 per cent of homeless people in the EU are migrants and refugees, and they are a particularly significant proportion of the homeless population in countries such as Italy and Greece which have experienced high levels of immigration (Harvey 1994). Immigrants from outside the EU appear more likely to become homeless than immigrants from other member states, and illegal immigrants face the greatest difficulties of all (Daly 1993).

Data from Italy suggests that homeless immigrants have a different profile from the rest of the homeless population. For instance, they were better educated than the Italian homeless, with 9.6 per cent having had higher education in comparison to only 4 per cent of homeless people in Italy overall. They were also more likely to be employed and enjoyed stronger social networks and better health than other homeless people (Tosi 1995).

Rights to housing in the EU

FEANTSA argues that the most effective way to address the problems of homeless people in the EU is to adopt a rights-based approach, and therefore they have conducted research into the rights of homeless people to housing in Europe.

The EU has no housing competence and no right to housing is enshrined in any of the European Treaties or in European legislation. Therefore homeless persons within the EU must rely on domestic law to establish their rights to accommodation.

Direct rights to housing exist only in France and the UK. An enforceable right to accommodation was introduced in the UK by the Housing (Homeless Persons) Act 1977. However this right only extends to certain 'priority' groups of homeless people, including families with children, pregnant women, victims of fire, flood or other such emergencies, and people who are vulnerable due to old age, mental illness or handicap, physical disability or other special reason. The UK government has now abolished the right of homeless people in England and Wales to permanent rehousing. Instead, they are entitled to temporary accommodation for a limited period of two years, although under certain conditions this entitlement may recur for further periods. Homeless households must now be assessed alongside all other applicants in a single waiting list for secure local authority housing. There have been no changes thus far to the legislation in Scotland and Northern Ireland.

In France a right to housing was introduced by legislation which came into force in 1990 (Drake 1994). It is called the *Loi Besson* (after the housing minister at the time) and requires local authorities to make a plan for housing all disadvantaged people in their area (including the homeless), and provides earmarked funds for implementation of the plan through partnerships of public authorities, private landlords and housing associations. Harvey (1993) argues that the *Loi Besson* represents an attempt to realise a right to housing by tying it to a package of planning and finance. However, whilst this measure refers to housing as a right for all citizens, it does not offer an enforceable legal right to accommodation for any particular individual or household.

In a number of states an indirect right to housing could be said to exist by virtue of legislation and/or practice including Denmark, Belgium, Germany, Luxembourg and the Netherlands. In these countries public welfare law stipulates that homelessness is a social ill to be avoided, and local or regional authorities are funded to take action to prevent homelessness or to assist those who become homeless. In each of these states legal instruments define criteria for preferential access to social housing for homeless people, but these are enabling provisions which do not entitle homeless people to accommodation (Avramov 1995). In Belgium a right to housing was enacted in the new constitution which came into force in February 1994 (Daly 1994).

The constitutions of both Spain and Portugal embody a right to housing, however no mechanisms exist in either country whereby citizens can realise this right. These constitutional provisions have thus far been of little practical consequence.

There are no direct, indirect or constitutional rights to housing in Greece, Italy or Ireland. The Housing Act 1988 in Ireland empowered local authorities to provide accommodation to homeless people, but it did not oblige them to do so except in relation to homeless children up to 18 years old (Daly 1994). In Italy there is a general right to services of the state and in some cities (mainly in the north) municipal or regional authorities do provide housing to the homeless. Legislation in Greece sets minimum standards of decent housing, but the state is not obliged to ensure that people have access to housing which meets those standards.

In most member states such housing rights as do exist are only available to citizens rather than residents, and thus immigrants are excluded. Also, obstacles such as residency or registration requirements in some countries, including Luxembourg, prevent people of 'no fixed abode' exercising their rights. Furthermore, the practical administration of these housing rights varies enormously within countries, particularly in member states such as Germany with strong regional government.

Services for homeless people in the EU

FEANTSA has reviewed the map of rights and services available to homeless people in the EU and has discerned a broad north/south split with countries such as Denmark, France, Germany, Belgium and Luxembourg providing the best services to homeless people, and Spain, Portugal and Greece the worst (Daly 1992; Drake 1994). I shall now outline briefly the types of organisations working with the homeless in the EU, and the sorts of services they provide.

Most services for homeless people in the EU are provided by voluntary organisations rather than the state. Whilst the ratio is two-thirds private/one-third public throughout the EU, there are considerable variations between countries (Daly 1992). For instance Great Britain, and particularly Scotland, has a high level of public sector involvement in homelessness services, as have Greece, Spain and Portugal. On the other hand, over 90 per cent of organisations working with the homeless in Luxembourg, Belgium and Ireland are from the voluntary sector. In a number of countries, including Germany, Italy, Spain and Ireland, religious organisations are the main source of this voluntary sector provision. The predominance of 'private' rather than 'public' provision of homelessness services serves to highlight the failure of the state in many countries to care for homeless people, and also perhaps an attitude that homelessness belongs in the sphere of charity rather than entitlement.

Much of the service provision for homeless people in Europe is aimed at meeting basic needs such as shelter, food and clothing. These services were provided by an average of 80 per cent of organisations in each country surveyed by FEANTSA (Daly 1992). Information to the homeless about matters such as their rights and the availability of accommodation was the next most common aspect of service provision, supplied by over 60 per cent of organisations surveyed. Services intended to promote the long-term re-integration of homeless people into the labour market and/or a settled way of life were less widespread, with only around half of the organisations providing this sort of support. Re-integrative services were most common in Denmark, Luxembourg, France and Belgium, and least common in Spain, Greece and Ireland.

Conclusion: the role of the EU in homelessness

This review of homelessness in Europe has several important implications for the future involvement of the EU in this domain of social policy. First, it is clear that the housing market is a key element in the generation of homelessness, but also that homelessness is a multi-dimensional phenomenon inextricably linked to wider processes of social exclusion in European society. The EU therefore has a role to play in combating homelessness as part of its social cohesion agenda, whether or not it gains competence in the housing field.

Second, it is clear that many common processes and factors underpin the growth in homelessness across the EU. However the character of the phenomenon still seems to differ significantly between countries so that an EU-wide policy on homelessness may not be helpful. It is important that responses to homelessness are sensitive to varying local conditions and thus national and regional levels of government are best placed to devise general policies.

Third, greater access to EU Structural Funds for projects working with homeless people, particularly programmes intended to assist them to re-integrate into mainstream society, would be helpful. These types of programmes are underdeveloped across much of Europe and are highly relevant to the EU's social cohesion agenda. Funds spent on homeless people should not be limited to their education, training and employment needs, but should also provide for their rehousing as accommodation is the most pressing need of many homeless people.

Note

1 This chapter draws heavily on national reports and overview papers produced by the European Observatory on Homelessness, which was established in 1991 and is managed by the European Federation of National Organisations Working with the Homeless (FEANTSA). As I have had to depend on the report of the Observatory national correspondent for most countries, these observations are usually based on the description and opinions of one author per member state.

References

Avramov, D. (1995) *Homelessness in the European Union: Social and Legal Context of Housing Exclusion in the 1990s: Fourth Research Report of the European Observatory on Homelessness.* Brussels: FEANTSA.

Avramov, D. (1996) *The Invisible Hand of the Housing Market: A Study of Effects of Changes in the Housing Market on Homelessness in the European Union: Fifth Report of the European Observatory on Homelessness.* Brussels: FEANTSA.

Daly, M. (1992) *European Homelessness – The Rising Tide: The First Report of the European Observatory on Homelessness, 1992.* Brussels: FEANTSA.

Daly, M. (1993) *Abandoned: Profile of Europe's Homeless People: The Second Report of the European Observatory on Homelessness, 1993*. Brussels: FEANTSA.

Daly, M. (1994) *The Right to a Home, the Right to a Future: Third Report of the European Observatory on Homelessness, 1994*. Brussels: FEANTSA.

Drake, M. (1994) *Homeless People in Europe and Their Rights*. Brussels: FEANTSA.

Eurostat Yearbook (1995) *A Statistical Eye on Europe 1983–1993*. Luxembourg: OOPEC.

Eurostat Yearbook (1996) *A Statistical Eye on Europe 1985–1995*. Luxembourg: OOPEC.

Eurostat Statistics in Focus (1995) *Population and Social Conditions: International Migration in EU Member States – 1992*. Luxembourg: OOPEC.

Feijter, H. and Radstaak, H. (1994) *Homelessness in the Netherlands*. Brussels: FEANTSA.

Harvey, B. (1993) 'Homelessness in Europe – National Housing Policies and Legal Rights', *Scandinavian Housing & Planning Research* 10: 115–119.

Harvey, B. (1994) 'Homelessness in Europe', speech to European Network for Housing Research conference, University of Glasgow, 29 Aug.–1 Sept.

House of Lords Select Committee on the European Communities (1994) *The Poverty Programme*. London: HMSO.

Maclennan, D., Stephens, M. and Kemp, P. (1997) *Housing Policy in the EU Member States: Report to the European Parliament*. Luxembourg: European Parliament Directorate General for Research.

McCrone, G. and Stephens, M. (1995) *Housing Policy in Britain and Europe*. London: UCL Press.

Neves, V. (1995) Unpublished notes on housing in Portugal.

O'Sullivan, E. (1994) *Homelessness, Housing Policy and Exclusion in the Republic of Ireland*. Brussels: FEANTSA.

Rostgaard, T., Jenson, M.K. and Koch-Nielsen, I. (1995) *Homelessness and Social Exclusion in Denmark*. Brussels: FEANTSA.

Sapounakis, A. (1994) *Annual Report on Homelessness in Greece*. Brussels: FEANTSA.

Specht-Kittler, T. (1994) *Housing Poverty in a Rich Society: Houselessness and Unacceptable Housing Conditions in Germany*. Brussels: FEANTSA.

Stephens, M., Bennett, A. and Smith, F. (1997) *Housing Associations and the European Structural Funds in Scotland*. Edinburgh: Scottish Homes.

Tosi, A. (1995) *Homelessness in Italy*. Brussels: FEANTSA.

Webb, S. (1994) *My Address is Not My Home: Hidden Homelessness and Single Women in Scotland*. Edinburgh: Scottish Council for Single Homeless.

12

MARGINAL HOUSING ESTATES IN EUROPE

Anne Power

The European Union is reaching far deeper into the lives of its 300 million citizens, particularly its poorer citizens, than is immediately apparent. Right across the Continent, governments are striving to join the European monetary union in 1999 with all its penalties, its Deutschmark domination and its unpopular limitations on the economic power of individual governments. Many expect the benefits in wealth and work creation to outweigh the costs but, in the short term, the process is painful. As a result of it, government infrastructure and support is under its strongest challenge for generations. People living at the very bottom of society in marginal areas, on marginal incomes or on government subsistence, are increasingly squeezed by the lowering of real wages, the up-skilling of the job market, the consequent reduction in jobs for them, and the reductions in services that accompany the general tightening of public belts. This makes publicly funded housing estates and other public projects vulnerable to many negative political, social and economic pressures, hitting hardest, where it hurts the most, among the very poor. This chapter describes how marginal housing estates came to house the poorest groups, how they are affected by wider pressures and why they are proving an undervalued but essential resource in the period of rapid transition in Europe at the end of the millennium.[1]

The study of twenty unpopular European estates set out to uncover the nature of estates, their social and organisational characteristics, the causes of their decline and attempted rescue, the prospects for their future viability. The five countries, France, Germany, Denmark, Ireland, Britain, were chosen because of their geographic links, the strong public involvement in housing and cities, the varied, but nonetheless common, commitment of their governments to social and urban underpinning, their common experience of post-war mass housing construction, its rapid decline and the attempted rescue of the most marginal estates. The experience of mass housing could serve as a model for other essential services in terms of both the negative pressures and the scope for change.

Large modern publicly funded estates were first built as Utopian solutions to urban housing problems; they are now widely regarded as unpopular and conspicuous failures of an over-rigid, over-ambitious, over-paternal and arrogant system of government. They also epitomise the problems of transition from old-style hierarchical and universalist government provision to new-style fluid, varied, enabling government roles. Mass estates are no longer valued for their success, rather they are used as a housing safety valve to cope with the pressures on marginal groups. They provide a useful example of essential change because they demonstrate the role of government underpinning in their survival, the potential for adaptation and innovation in their new roles, the significance of targeted effort and local autonomy in their rescue, the dangers of over-prescriptive, uniform solutions to urban problems (Ministère de la Ville 1993). Most importantly they illustrate both the inevitable pressures on marginal communities and a possible alternative to American urban outcomes – massive ghettoisation and abandonment of cities.

The problem of mass housing

In the period of chronic shortage following the Second World War, the five countries examined here faced widespread housing deficits running into many millions of homes. In France and Germany alone, at least 26 million additional homes were needed at the end of the war. Over the first post-war generation, politicians made their mark in two main ways: breaking with the past and adopting large-scale solutions to shortages. Nowhere was this more true than in housing, and nowhere was the potential for political success more obvious. Massive bomb-damage, huge movements of refugees, uncontrolled influxes from countryside to towns and rapid industrial expansion all created the momentum to build on an unprecedented scale, using new techniques based on the mass-production, factory model. Politicians seized upon the modernist idea of ultra modern, cellular, pre-cast homes in giant high-rise blocks as a breakthrough in futuristic imagination (Quilliot and Guerrand 1989). Popular support for units built depended mainly on volume, and new ideas about 'streets in the sky' and 'machine living' had a liberating tinge. The idea of obliterating squalid old slums and creating a uniform, replicable, neatly packaged solution to a baby-boom, homelessness, squatting and refugee camps was irresistible. Modern homes with new fittings and unprecedented amenities were a large leap forward from previous urban conditions let alone the recent upheavals and destruction of war. The fact of massive urban dislocation was seen as a positive break with the failed past rather than a negative disruption of tried and tested techniques of expansion.

In all, over twenty-five years from 1950 to 1975, around 15 million new publicly funded housing units were built in the five countries, many of them on large estates, usually on the edge of existing towns and cities, usually in concrete, using industrial techniques, often in high-rise blocks of more than five storeys, invariably utilitarian, monochrome, imposing in style (Vestergaard 1993; Levy 1989; BRBS 1986; Burbidge et al. 1981).

Although the units were expensive to build – possibly more costly in the long run than conventional housing – the gigantesque style was driven by political fantasy about scale solutions and by building and architectural fashion (Dunleavy 1981). Public funding seemed a price worth paying for the obvious gains of scale. Few questions were asked, certainly not of prospective consumers, and political reputations were made or broken by the ability to deliver a quarter of a million units a year or more – an approximate goal sometimes met in Britain and frequently exceeded in Germany. In France, where a strong central state determined urban policy, the goal was to implant estates of 5,000–10,000 units in semi-rural communes outside major cities, built using specially laid railway tracks to carry the huge cranes that were needed to construct high, closely packed blocks, thus facilitating dense forests of towers up to twenty-five storeys high. The whole idea of estate implantations in and around cities was based on a rolling programme that would gather momentum, producing falling costs and shrinking time scales as the units and programmes multiplied, flattening opposition and forcing divergent communities into a single mould, based on machine structures (Quilliot and Guerrand 1989).

The juggernaut image of rapid and runaway change, used by Anthony Giddens in *The Consequences of Modernity* (1993) fits the mass housing model, for the construction of estates galloped ahead at a pace that far outstripped organisational or political control or customer feedback.

Landlords had little experience of the management of industrially constructed mass estates on which to draw, thus all parties to the development were largely unprepared for the consequences. The effect of mass housing on potential opposition within urban communities has been likened to a steamroller, for programmes, once agreed, could not easily be deflected from the set path, rather they flattened opposition reducing ideas about strategic housing solutions to a uniform mould, shaping future patterns through compacting into hard, even forms previously jumbled and uneven patterns of development.

Most of the largest and clumsiest mass estates were built between 1960 and 1975, although many were planned earlier and some construction continued to the early 1980s. Initially, there was no shortage of demand for new homes due to the build up of pressures on fast-growing urban economies. But, over a relatively short period, mass housing became undesirable to families with children, who preferred more conventional locations nearer to jobs and more conventional dwellings nearer to the ground. Streets in the sky embodied fatal misconceptions about human interaction and social control that we uncovered in all countries. About one-third of social housing, often in the biggest and most modern estates, gradually became unpopular. Populations in estates began to shrink soon after the early occupation, and demand began to evaporate in all countries by the mid to late 1970s as more economically secure households exercised choice and either bought small houses in the booming private housing market, or opted for smaller-scale social housing or even private renting, as happened sometimes in Germany, France and Denmark, because it was often cheaper, more centrally located, more conventionally designed and managed (Power 1993).

The European urban management system of control by concierges and door porters proved invaluable in protecting the private rented stock in cities, and on the Continent this helped to sustain renting as the major urban tenure over a longer period than in Britain and Ireland where traditionally flats were more uncommon and therefore caretaking less accepted as intrinsic to rented housing. Although the continental system of caretaking was transposed to large, modern outer estates, all sorts of problems quickly overtook it, such as the vastly different scale of operation, the lack of conventional street life, the unsettled nature of the new communities and the remoteness of landlords.

Mass housing estates were difficult to manage because of their physical scale, the uniform patterns of use, based on a dormitory model of housing devoid of economic activity; and the communal layout of estates, militating against personal investment and stake in the area or family and household control over the immediate environment. Not only were the problems of enforcement complex but the whole pattern of informal social control which operates in more settled communities failed to take root in many mass estates. These problems dominated large public housing developments even where homes were conventionally designed, leading to a deadening dullness and lack of varied uses. This in turn had serious consequences for social life. Not only were incoming inhabitants often unused to the new physical structures and found them inhospitable, but the newness of the community meant that there were few recognised signposts or signals for communal behaviour in a futuristic and implanted environment.

It was hard to establish acceptable patterns of behaviour or to exercise control outside the front door as other people's behaviour was likely to impinge in negative ways, even where the behaviour itself was not unreasonable. Ordinary family interaction, domestic noise, child supervision, teenage group dynamics and so on were magnified many times over in mass estates because a single incident could have a strong impact on the hundreds of individuals or households who shared the common spaces. Young people gathering in stairwells or children playing on the grass are two examples. This dislocating interaction between families, generations and cultures created an unease among residents that was compounded by similar problems facing the landlords. To contain child nuisance, for example, many normal activities such as informal social gatherings on grassed areas would be curtailed.

Landlords and management

Landlords attempted to exercise control over conditions on behalf of residents but rarely had sufficient staff based at the front line to do this effectively, or sufficient awareness of the compound effects of scale, unfamiliarity and uprootedness to realise until much later the requirements of mass housing management. They did not know how to encourage, facilitate or impose new patterns of neighbour relations in a disconnected area since the speed and scale of mass housing were unprecedented. Neither was it possible for residents to identify with the surrounding community where communality was enforced and where informal communication patterns had had no time to emerge. Possibly the most serious consequence of the mass housing approach was the failure to foresee new community problems implanted in separate and often isolated areas, with insufficient organisational infrastructure, inadequate links with the city, few settling in or stabilising mechanisms and no public voice for the new residents. People's lives were atomised within 'machines for living' (Le Corbusier 1946).

Because of the lack of feedback, problems went undetected and multiplied rapidly. The very structure of large separate housing estates guaranteed a chaotic effect leading to a rapid tipping of conditions from new-found solution into recreated slum-like problems. For example, to take a very British example, if dogs were not allowed, landlords were expected to enforce the ban, but with what authority? The ultimate sanction was eviction but courts often refused to evict families for such reasons, even if it was a known transgression of housing rules, imposed for sound reasons related to public health and child safety. Families could appeal on grounds of fear of burglary or family attachment to an existing pet and the breach would itself generate further breaches to a point, rapidly reached, where dogs would wander on decks, tearing refuse bags, fouling pathways and creating a menacing environment (DoE 1981). The ban was based on the difficulty households faced in controlling dogs in multi-storey blocks, but the breach was based on past practice, the need for homely signs, and a fear of the new environment. Loss of control and greater fear were the outcome on many high-rise and mass estates.

The argument about banning dogs could be applied to almost any aspect of modernist estate living; car-parking and repair; gardens, play areas and the environment generally; lifts and entrances; refuse disposal; control of children and teenagers; repair of lighting, letter boxes, bells and other accessible, communal facilities. Each of these aspects of estate living required a high level of supervision and control; ready acceptance by all residents of house rules; quick, enforceable sanctions. In practice none of these three conditions was in place at the outset because politicians, builders and landlords were equally ignorant of the impact on community conditions of their imposing betterment plans. The result was the rapid decline over less than fifteen years of many estates from a heralded solution to a nightmare of disorganisation and disarray (BRBS 1988).

The common problems resulting from largely untried experiments in urban living disguise some major differences in approach. The continental landlords in the survey were non-profit housing associations, co-operatives or private landlords, rather than direct government bodies like councils in Britain and Ireland. Management styles varied between strong policies of enforcement which might ensure high quality caretaking and repair, but debar children from playing on grass or residents from organising parties or even meetings, and weak policies of enforcement which offered little supervision and left minor problems unrectified on the basis of which bigger problems quickly mounted. The former strong management approach held conditions for longer and was generally adopted by continental landlords, partly stemming from their more independent and private funding system, partly from their more mixed, more business-oriented housing systems and more robust housing management tradition (OECD 1987). The latter, weaker management approach was more commonly found with council landlords, partly resulting from their more bureaucratic and less responsive, more routinised public management style, partly from more pragmatic approaches to enforcement, partly from a weaker management tradition generally, particularly in ground-level caretaking (Burbidge *et al.* 1981). The result was much worse superficial conditions on estates in Britain and Ireland than in continental Europe and the much more rapid emergence of problems, signalled in the late 1960s by the spectacular explosion of Ronan Point Tower, owned by Newham Council in the East End, and the immediate scrapping of high-rise subsidies (Dunleavy 1981).

One lesson was clear from the early experience of mass housing estates; without intensive, highly localised supervision and management, dense, mass housing estates did not become hospitable social environments or even, in extreme cases, remain useable (BRBS 1991). While management is pivotal to the success of any business, housing management provides the axis for social rented housing without which tenants cannot enjoy basic domestic security or a safely maintained home environment. This is even more true of mass housing due to the complex building form, social instability, and intensive maintenance requirements. The fact that rented housing is a long-term investment and business was often overlooked in the over-hasty development focus. Under both the continental non-profit and British or Irish council structures, management systems were not sufficiently robust to withstand wider pressures on marginal communities, although our study highlighted the differences in management impact (Elie *et al.* 1989).

Polarisation and race

A direct consequence of the early decline of mass housing was its loss of popularity among better-off workers, but its emerging importance to poorer people, including large foreign immigrant communities that had come to Europe during the post-war boom, an era of full employment. Over less than a generation many of the expanding low-skilled jobs migrants came to fill began to disappear from building, heavy industry, textiles and public services. By the late 1970s it was often far harder for the children of immigrants to find jobs than it had been for their parents as new arrivals. This shift in employment away from manual and low-skilled work affected minorities disproportionately but it also decimated the economic base of traditional white communities in older industrial areas such as Glasgow, Lancashire, the Ruhr, Pas de Calais and Alsace.

Better qualified and more enterprising younger households left the declining areas, causing a shift in demographic patterns by 1980 that had been almost completely unforeseen. The effect on mass housing in the collapsed industrial areas, and in the less successful neighbourhoods of metropolitan cities, was catastrophic, leaving marooned communities with insufficient work and leaving social landlords with under-used assets and deeply marginalised communities. At the bottom of the hierarchy of areas, neighbourhoods and estates in each country, was a pool of marginal housing that was difficult to let and difficult to manage. The vacuum created by this slump in popularity attracted new problems.

The migration from overseas created particular tensions in European cities as around 10 million new workers and their families were absorbed from exporting countries around the globe. A major transition in populations happened over a relatively short period in three phases. Crowded, older inner city areas housed growing numbers of new arrivals, as established residents moved to the new estates and more suburban developments. The newcomers came under increasing pressure as urban renewal and gentrification in the wake of mass housing began to displace the poorer households. Gradually households of foreign origin found their way into social housing, particularly the large and relatively new but unpopular estates attached to major cities (Gibbins 1988).

Thus race became linked to the poorest and least popular housing areas, both in inner city areas and in the new estates. This did immense harm to new and vulnerable communities because racially distinct minorities became identified with steep housing decline, as they replaced the native-born population first in inner areas, then in some of the most unpopular estates, as these areas became unacceptably run-down and unpopular. When this happened in older inner city neighbourhoods it was easier for people to blend into the landscape as streets and communities were closely juxtaposed, even though declining city neighbourhoods from Berlin to Paris, Copenhagen to London, often became labelled by their minority populations. But when minority groups moved into large, conspicuously separate and declining modern estates, their very access to estates signalled a watershed in their unpopularity, often precipitating further decline as other groups, with more choice, refused to move in, or moved out even faster as a result of the visible lettings difficulties. Quickly some of the most stigmatised estates shifted from housing almost exclusively European, employed populations to housing high proportions of ethnic minorities doing menial work for low wages in an economic situation where work was increasingly scarce. Through this process many estates became more racially mixed, poorer, more clearly labelled as such and therefore more poorly serviced and more isolated (Provan 1993).

No one has explained precisely why the poverty and low status of certain communities invite a lower level of public service. It relates to the inability of service users to have leverage over service providers because of their perceived inferiority. It also derives from the limited purchasing power of residents and the greater obstacles facing lower income households in maintaining standards. The attitude of beleaguered managers towards needy tenants who place additional demands on their time and energy can become extremely negative. Human beings, including providers of public and social services, have a tendency to generalise from conspicuously bad experiences of human problems, leading to the stigmatisation of whole groups, normally those groups who are recognisably different. Staff often respond to negative reputations and signals by a lowering of effort in a situation where additional effort is required to maintain conditions and reduce the gaps that constantly open and re-open in fast-changing conditions (Reich 1993). All these factors, and many others, accumulate in relation to certain areas that are more difficult than average to manage, creating poor standards of basic repair, cleanliness and environmental maintenance, leading to visually segregated environments.

It is an idea strongly rooted in Scandinavian urban policy that environmental signals can determine the status of areas and therefore by inference of residents (McGuire 1981). The ranking of neighbourhood environments deters more successful residents from and encourages more disadvantaged and excluded households to poor areas. Over time therefore, on estates, many different and subtle forms of decay applied downward pressure on lettings and on management performance, making it ever harder to hold conditions and deterring ever more strongly the economically more successful applicants. All five governments by 1975 had begun to experience spiralling conditions on estates, leading them to recognise the flaws in the mass housing approach and to adopt different approaches to housing and community needs.

International survey of mass housing estates

The study of mass housing problems began in earnest in Britain in the mid-1970s when the Greater London Council was persuaded by Islington residents to abandon plans for yet another mass housing estate, in favour of a park, due to the dearth of demand for new flats and the high refusal rate for offers of nearby new council housing (North Islington Housing Rights Project 1976). Around the same time Glasgow City Council started to demolish an unfinished estate because of the collapse in demand and Liverpool offered an empty ten-year-old tower block for sale for only £1 (DoE 1981). Between 1976 and 1979, the British government's inquiry into difficult-to-let estates confirmed that the problem was nation-wide, included old and new estates, houses as well as flats, London as well as the North. It affected up to half a million units, but possibly three times that number (Burbidge et al. 1981).

In an attempt to uncover whether different ownership and management patterns affected the extent of the problems of mass estates, conditions in four other European countries were examined between 1987 and 1995 on the basis of visits, interviews with key representatives and scrutiny of research being conducted within those countries. It was quickly discovered that, to a greater or lesser extent, all five countries now experienced similar problems. Research into the problems demonstrated that the decline was occurring on a large scale, was causing serious social consequences and at least in part was the result of neglect and mismanagement both of the buildings and the environment.

Between 15 per cent and 30 per cent of the social rented stock in each country was affected. At least 2,000 estates comprising 4 million dwellings were being targeted by government programmes to rectify problems (Power 1993). The scale of the problem appeared greater in Britain and France than in Germany, Denmark and Ireland, though the eastern regions of re-unified Germany had mass housing problems of a totally different order of magnitude from elsewhere in the study, with 125 giant estates of over 5,000 units. These were not included among the twenty survey estates.

There was a remarkable coincidence of experience and policy making between the different countries. Between 1978 and 1987, the five different governments launched rescue programmes involving renewed government intervention, to restore physical, financial, organisational and social viability to mass estates. These programmes aimed to address a set of common problems within each country on the most severely affected estates. As far as could be uncovered, there was little exchange of information or experience between governments and landlords across national boundaries before the onset of the crisis and the launch of rescue initiatives although the pattern of mass estate development had been broadly copied and actively marketed by volume builders and urban design companies. The rush of governments towards a flawed construction model accompanied by the lack of dialogue or debate about its possible consequences made the pattern of intervention in the rescue of estates and the findings of the survey more interesting and their impact more convincing.

Governments intervened at the level of local estate communities because of the intensity of the localised problems, the political liabilities of mass housing failure, and the continuing pressure to house poor people to prevent homelessness. A decisive factor in the rescue attempts was the threat of disorder in volatile estates, a fear that became reality as riots occurred in Britain and France in 1981, after their early rescue programmes had begun (Dubedout 1983). The precarious financial position of social landlords attempting to hold conditions on large estates, particularly under the arm's-length continental model, generated a sense of urgency. This anxiety was intensified by the spectacular collapse of Neue Heimat, based in Hamburg, the largest social landlord in Europe, with nearly half a million units. The giant non-profit company and its many profit-making subsidiaries disintegrated in bankruptcy amid allegations of corruption, mismanagement and over-ambitious expansionism (Fulrich and Meuter 1987).

The twenty estates in this survey were drawn entirely from those included in the government rescue programmes. This was done in order to ensure that the survey estates reflected the most serious and most conspicuous core problems of each country, and were sufficiently problematic to warrant special attention, having already been selected according to government and research criteria within each country. These criteria were strikingly similar between countries, but the most salient were: physical scale and disrepair; serious lack of demand and unlettability; financial and management unviability; deteriorating social conditions; problems on a scale and of an intensity that required special external interventions. A further rationale for focusing the study on government rescue programmes was to be able to analyse and compare different government

223

approaches to mass housing problems and to assess the impact of rescue pro-grammes. In this way it might be possible to establish how potentially viable such estates are, how far their problems are attributable to physical or organisa-tional or social problems; also to explore which problems are susceptible to which solutions, and at what cost. Next we present a summary of the main findings.

Findings from twenty European estates

Table 12.1 outlines the main characteristics of the estates, the most significant of which were industrialised design, large scale and peripheral location. Table 12.2 summarises the problems most commonly found on the estates prior to the res-cue programmes leading to chronically decayed conditions. Governments and landlords in all countries confirmed the primacy of physical, social, organisational and financial issues. Problems were deeply interlocking, complex and multi-faceted. Most estates were severely affected by most problems. The physical scale interacted with management in a way that seriously undermined social condi-tions. For example, design often led to insecurity and damage which in turn

Table 12.1 Characteristics of twenty European estates

Characteristics	Condition	No. of estates
1. Location	Periphery	18
	Inner city	2
2. Transport	Poor	16
	Adequate	4
3. Landlords[a]	Public	8
	Publicly sponsored	2
	Privately sponsored	6
	Co-operative	1
	Private	1
4. Size of estate	Under 1,000	5
	1,000–5,000	8
	5,000+	7
Average size	3,600	
5. Construction	Industrial or semi-industrial	19
	Traditional	1
6. Structure	Mainly flats	16
	High rise[b] (5+ storeys)	14
	Mainly houses	4
	Only medium and low rise	6

Source: Author's visits and government research.

Notes
a   Several continental estates had more than one landlord; all public landlords were British or Irish.
b   Most high-rise estates included some medium- and low-rise blocks.

Table 12.2 Problems most commonly found in twenty estates

Problem	No. of estates
Design	
Communal environments	20
Weak security	20
Over-large blocks	19
Overall scale of estate	18
Structural faults in blocks	18
Communal problems	
Lack of privacy	20
Unsupervised common areas	18
Isolation of estate from urban landscape	18
Monotonous design and environment	18
Limited social contact due to communal problems	18
Management and financial problems	
Population turnover and instability	20
Maintenance costs and complexity	20
Insufficient local management organisation	20
Weak caretaking structures	20
Vandalism and anti-social behaviour	20
Lettings problems and vacancies	19
High management costs and arrears of rent	19
Settling in problems and lack of support	19
Extreme conditions	16
Disintegration of management control	10
Concentrated social disadvantages[a]	
Poverty	20
Concentrations of lone-parent families	19
Young people with low skills and without work	19
Crime and policing difficulties	18
Ethnic minority concentrations	11
Disorder	8

Source: Author's visits and interviews; government research.

Note
a   Defined as areas with higher than average concentrations of each disadvantage compared with surrounding areas and social housing overall.

created repair and policing problems, creating a high turnover of tenants and serious community instability. Though they ranged in severity, the overall effect across all five countries was to create communities facing the threat of incipient social breakdown.

Problems were extreme in sixteen estates, that is they had reached a point where landlords no longer believed they could maintain the viability of the estates. In four cases problems were serious but had not yet reached a point of threatened disintegration. In ten of the sixteen extreme cases, including the five case studies, chaotic conditions prevailed – that is, conditions in which lettings, rent collection, supervision and maintenance had effectively lapsed and the estates had become unmanageable.

An important finding to emerge from the analysis of problems was that ethnic minorities were not directly linked to extreme problems even though they were frequently blamed for the acute decline. This was demonstrated by the fact that many of the most difficult estates, particularly in Britain and in Ireland, had virtually no ethnic minority populations. The most extreme cases in Ireland experienced as serious physical, social and economic problems as those in the rest of the survey, coupled with more severe management problems, in spite of the homogeneous populations and relatively integrated cultural patterns. The acute decline of estates housing virtually all-white communities demonstrates the over-riding influence of wider economic factors over the more easily blamed social and cultural factors which were found to be a consequence rather than a cause of decline (Power 1993; Wilson 1996). In spite of this, it was true that in cities with high concentrations of ethnic minority communities, they were often highly concentrated in the most unpopular estates.

Table 12.3 shows the concentration of need on the five case study estates. It also shows the extremely high levels of unemployment which often affected racial

Table 12.3 Concentration of ethnic minority populations and levels of unemployment in five case study estates

	France	*Germany*	*Britain*	*Denmark*	*Ireland*
Proportion of ethnic minority households	30% (of whom ⅔ North African)	70% (1,755 Turkish, 1,022 other foreign, i.e. ⅔ Turkish, ⅓ Germans, Poles and so on	60% (primarily Afro-Caribbean – 42%)	25% of units/ households (40% of population, of whom vast majority are Turkish)	Virtually none
Levels of unemployment	13% but 37% for foreigners (4.5% in 1975)	30%	40% 60% among young	15% officially (increased 5-fold 1970–85 – 45% of all adult households have no earner[a]	60% (of whom 80% long term, i.e. over 2 years)

Source: Windsor Workshop (1991); follow-up information from estate visits and from Census.

Note

a Including many cases of early retirement.

minorities disproportionately, but was extraordinarily high in Ireland where race was not a factor (Peach 1995). In other words, where migrants had been drawn in to fill the low-skilled jobs of the earlier growth periods they were disproportionately affected by the changes and over-concentrated in the most problematic areas (Holmans 1995). Where immigration had not occurred on a large scale, in Ireland and in other regions suffering long-term decline, low-skilled indigenous populations experienced similar pressures. Race was not the central determinant of disadvantage.

Breakdown

From the mid-1970s onwards, large urban populations were affected by new and extremely rapid changes, some triggered, some accelerated by the oil crisis. The mass estates, unpopular as they had often become among households with choice, were an invaluable housing resource that was both indispensable and irreplaceable in the pressured economic climate of the late 1970s and early 1980s. Many low income and newly formed households faced increasing difficulty in getting on the first rung of the housing ladder. Mass estates often provided quick access because of low demand. While the change in housing role of the estates in favour of ever poorer households had encouraged an exodus of better-off households, that in turn created additional space for excluded groups. The resulting stigma fuelled further exodus, creating such strong downward pressure that some estates became up to one-third empty.

By an irony of unintended consequences, letting almost exclusively to the poorest groups on the basis of need, and slack demand, generated such extreme social instability that even very needy households would sometimes refuse offers of accommodation in the survey estates. As a result the volume of empty property and the pace of population turnover rose steeply. Table 12.4 shows the fluctuation in population, the scale of turnover, and the level of empty property in the five case studies during the period of acute decline and rescue. The problem of collapsing demand not only had serious financial consequences but devastating social consequences too. Signs of abandonment invited criminal damage, theft of fittings, invasion by gangs and criminal networks, in many cases clearly linked to the drug trade. The process of exodus then became self-fuelling and existing management structures proved completely unequal to the task, with staff resorting to fire-fighting, extreme security measures and even more desperate lettings to vulnerable and needy households that simply could not cope with the pressures of high-rise living, inadequate services, and community instability in a desperate attempt to keep properties occupied. In the wake of such instability crime, violence, intimidation and fear rose to extraordinary levels up to eight times the average for surrounding areas (City of Cologne 1989; Downes 1991).

The breakdown of order was widely feared and signs of this happening were visible in most estates. Actual disorders had broken out in eight of the estates. Disorder was not confined to estates with large ethnic minority populations,

Table 12.4a Volume of empty units and level of annual turnover of tenancies in five case studies at the point of extreme decline

	France	Germany	Britain	Denmark	Ireland
Level of empty units as % of total	28	20	10	16	16
% population leaving the estate annually	30	25	20	35	30
Changes in size of population	35,000 (1975) 20,000 (1981) 25,000 (1991)	2,400 (1987) 4,000 (1991)	3,000 (1970) 2,000 (1983) 3,000 (1991)	2,200 (1985 estimate) 2,800 (1991)	15,000 (1970) 11,000 (1987 estimate) 13,000 (1991)

Source: Author's visits and Windsor Workshop (1991).

Note
These figures reflect conditions at the time of the most extreme decline.

Table 12.4b Improved performance after estate rescue programmes had been implemented

	France	Germany	Britain	Denmark	Ireland
Level of empty units as % of total	4 (1996)	0 (1995)	1.5 (1996)	0 (1993)	0 (1995)
% population leaving the estate annually	15	below 10	10	15	15

Source: Windsor Workshop (1991) and follow-up information from case study areas.

although much of the European debate about riots and disorder is framed as though this was the case (Mitterrand 1993). It also involved predominantly indigenous populations (Power and Tunstall 1997). However, a majority of estates, while assailed by multiple problems, had not suffered the extremes of street violence and police clashes. Nonetheless, in at least ten areas the hold of landlords or other authorities on conditions had reached a point of virtual collapse. It was this that led governments to intervene in support of new-style management initiatives (Elie *et al.* 1989). For these estates, found in France,

Germany, Denmark, Britain and Ireland, still housed large communities of several thousand people each, who needed somewhere to live. The estates were playing a precarious, but nonetheless vital, social role.

Government role in stabilisation

So far we have shown how estates were used to relieve some of the major housing problems facing cities, and the consequences for those communities of concentrating need in this particular way. Big changes in society such as job losses, family breakdown and migration were beyond landlord control, and in practice beyond the control of individual governments, but by pushing the most vulnerable and insecure groups into the most difficult environments landlords undermined the potential to help those very groups. The intensity of the pressures was new to social landlords and managing the pressures effectively had eluded landlords everywhere. The structure of support for estates had to change if landlords were to take the strain of greater social pressures (Ministère de la Ville 1993). The growing size of the groups that were most heavily dependent on subsidised housing, particularly those dependent on some form of income support, made the method of estate rescue even more important to urban stability, as the process and approach had as great a role in success as any physical changes in restoring viability. The outbreak of riots underlined this fact as the triggers were invariably linked to methods of control and organisation as much as other factors (Power and Tunstall 1997).

Governments everywhere believed that estate rescue and stabilisation were economically and socially preferable to renewed demolition in spite of the physical problems (BRBS 1990; SBI 1993). Unstable estate communities needed anchoring, not displacing, and available, affordable housing was needed to house the growing ranks of low income and workless households. Therefore, total demolition of estates was not on the renewal agenda although the removal of limited numbers of blighted blocks was undertaken in six estates. In spite of the basic decision to save the estates, there were at least two cases in Britain and Ireland, and one in France, where the survival of the estates is still in question (DoE Ireland 1997; Conrad 1992; DoE 1995).

In general, there was no question of governments funding replacement building programmes on a similar scale again and new social housing which was being built in the early and mid-1990s was on a much smaller scale and was many times more costly per unit than the rescue programmes in almost all instances. On the Continent, new funding was generally tied to small developments under 200 units (Salicath 1987; Power 1996; Power 1997). As a result most of the mass housing stock was being restored.

The approach to estate rescue was complex and in some ways circular as landlords aimed to attract more diverse populations to improve stability. At the same time they had to fill empty units to become solvent but they could no longer let to all-comers if they were to gain the support of more stable existing households whom they wanted to retain. They had to remedy physical and environmental defects in order to attract tenants but they needed additional resources to do this which would eventually come from additional rent income. They also had to involve and win the co-operation of existing residents in re-establishing basic control over conditions before they could start to make progress in their wider strategy of creating more mixed and therefore more socially stable communities. Much of their management difficulty stemmed from these conflicting requirements. Breaking out of the vicious circle was slow and complicated. It invariably required the injection of external resources. This is where governments stepped in.

Governments, in particular city governments, realised the imperative to find new and experimental approaches under the sheer pressure of estate decline which brought in its wake electoral problems. Social instability, high costs and unusable housing units provided a devastating and highly visible criticism of political failure. Therefore governments became central actors alongside landlords in a new approach to estate problems and estate renewal. Governments supported a multi-faceted approach to estate problems because successful rescue depended on highly localised responses to the actual conditions which, for all their broad similarities, varied in almost every aspect of detail.

Localisation

It became obvious through research that ground-level problems such as cleansing, maintenance, vandalism and lettings could only be tackled at ground level. Landlords recognised this same imperative which, obvious as it seems, had escaped the organisational imagination of estate creators and owners until it was almost too late. One obvious lesson of estate failure, accepted by all countries, was the need for a greater front-line presence, both to regain control over conditions and to open up lines of communication with residents and with the many non-housing services operating in relation to estates.

Governments funded remedial repair, environmental improvements, building upgrading and enhanced security, while the essential organisational reform was left to the landlords. Funding was invariably tied to resident liaison and improved management, requiring a local, team-based approach and strong social skills. The five governments funded social and training programmes, although only in France was this aspect targeted with government funding directly throughout the rescue initiatives (Power 1997). The new approach comprised structured consultation with residents to tailor improvements to actual experience; localised management of services to ensure focused performance and delivery; stabilised finances through targeted resources; modified rents and incentives; careful control of income, expenditure and conditions; low-level physical changes rather than radical replacement to encourage management viability; careful marketing of the estates and screening of incoming applicants to ensure social viability; a wide-ranging interconnected and interactive programme of changes.

Rescue programmes were housing-led because landlords, as owners of the estates, were the organisational axis for improvements. But in order to succeed they needed other services, such as education, health, police, as partners, and they needed targeted government resources to compensate for the concentration of need, particularly poverty and unemployment. As a result of the organisational shift to estates as management entities, resources were directed towards individual estates, both in cash for improvements and, as importantly, in people with skills to tackle management problems creatively. Every successful rescue project was developed with a locally based team that interacted directly with the estate community. The well-established management technique of focusing effort on troubled areas and releasing responsive managers from bureaucratic constraints to take on problems directly was put in place. Table 12.5 shows the types of

Table 12.5 Remedial measures applied on twenty European estates through rescue programmes

Improvement measures	No. of estates
Physical	
Building upgrading	20
Enhanced security	20
Environmental upgrading	20
Restoration of empty units	19
Decoration and enhancement of blocks	18
New or improved estate facilities	17
Dwelling alterations	14
Organisational/management	
Special estate-based rescue initiative	20
Intensified repair and maintenance	19
Localised landlord service, local team and manager	18
Lettings control to improve mix and stability	16
Screening of disruptive tenants	16
More intensive caretaking, cleaning and supervision	15
Financial	
More targeted spending through local management system	20
More rent income through greater control over lettings and higher localised inputs, higher lettings and lower turnover generating greater income	18
Greater viability through more diverse uses	18
Social	
Resident consultation	20
Additional projects for children and youth	19
Special support for vulnerable tenants	19
Resident representation on consultative and decision-making bodies	17
Employment and training initiatives	14
Ethnic minority support initiatives	11

Source: Author's visits and Windsor Workshop (1991).

measures introduced across the five countries, and the number of estates where each measure was found.

There was a remarkably coherent pattern of changes at estate level; greater local control over basic decisions affecting estates such as lettings; more responsive local services such as repairs and caretaking; environmental upgrading to create conditions that were more equal with other communities; social supports to compensate for concentrated social disadvantage; stronger links with near-by areas and the rest of the city. The links with the city would both help to create and result from more equal conditions, a symbiotic relationship that required carefully executed, tailor-made steps and sustained inputs. It would encourage a broader population base that was more integrated with the wider area. The urban linkage resulted directly from estate rescue programmes operating at three distinct levels: government at central, regional and local level; landlords at city, area and estate level; and residents at estate, block and small neighbourhood level.

Examples of change

When once a targeted initiative had been established on a particular estate, leading change agents were placed at the estate level (Elie *et al.* 1989; BRBS 1991). This immediately led to more localised jobs, more social and community facilities, more resident initiatives, more commitment to estates *per se*. Independent social landlords on the Continent encouraged local authorities to develop a stronger enabling role, while in Britain successful estate initiatives led to a radical decentralisation to estate level of top-heavy council services (Power and Tunstall 1995). As a result the number of empty units plummeted in the case study areas from a high of between 10 per cent and 30 per cent, to virtually full occupancy in all cases as reinvestment took place. Arrears fell as occupancy levels and population stability rose; repairs increased and conditions became more normal as reinvestment took place and rent income increased to more typical levels. Table 12.4b shows this improved performance in the five case studies as even extremely precarious estates became viable again. The swift transformation in conditions convinced residents, staff and politicians that positive change was possible.

In France, the change in approach was summed up by the slogan 'dropping the dinosaur image of social housing' (Quilliot and Guerrand 1989). In Germany, government researchers argued that estates which had been designed as cast-iron cities needed to be humanised with flexible, local management and resident involvement (BRBS 1991). In Denmark, managers moved out to estates to run environmental projects, building upgrading, lettings and transfers at ground level. They made their intensive custodial caretaking even more localised than this. In Ireland, council landlords sat in small flats with residents on estates, planning and replanning the changes. They helped train tenants' representatives and allowed tenants an input into lettings. In Britain, the bureaucratic rules surrounding council housing were suspended in experimental estates and many small-scale front-line initiatives took off that had a transforming effect on conditions.

There were many examples of the application of the principles of localisation through multi-faceted programmes coupled with linkage to the wider community. For example, the French case study estate handed over a whole tower to a local college. A lettings agency was set up to market the estate to young couples. An educational priority programme helped the ten primary schools on the estate to raise the reading age of children to the national average by setting reading targets for each class and providing special advisers for the estate schools (Elie *et al.* 1989).

The privately owned estate in Germany continued to house people in extreme need (almost entirely foreigners and ethnic Eastern Germans), but the independent management company that ran the privately owned estate exercised tight control over lettings, repairs, cleaning, security, investing in many more locally based staff and more intensive services from the additional rent income from occupied flats and a re-investment partnership supported by the City of Cologne (Hillebrand 1988). The company involved young people in training schemes to build playgrounds, a nursery and other amenities. The number of locally based repairs, cleaning and custodial staff shot up to over twenty for 2,000 units (Windsor Workshop 1991).

In Denmark, an estate of 1,000 units was divided into four caretaking and environmental areas, each with its senior caretaker, resident committee and local office, resulting in sparkling conditions and much higher resident satisfaction (Power 1997). The four areas of the estate developed their own plans for the open spaces with their own improvement budgets. The environmental work involved training and employing young people from the estate. Turkish residents from a rural background took the initiative in several of the gardening projects. The estates continued to house low income people but the commitment, the improvements and the local liaison changed actual conditions visibly (SBI 1993).

As part of these changes, the role of caretakers was upgraded in all the continental initiatives to enhance their control over conditions, while the front-line management back-up was greatly expanded and liaison with tenants emphasised (AKB 1988; SCIC Gestion 1995). It was a fundamental weakness of the British and Irish approaches to estate renewal that, in most areas, caretaking generally retained its low status and therefore estate conditions proved harder to restore (Power 1997).

Patchwork model

Each estate rescue programme, in order to succeed, based its structure on a combination of elements that emerged organically from obvious ground level problems, including local management, human resources, resident representation, targeted finance, re-investment, enhanced social facilities, services and activities, a focus on children and young people, training programmes for residents and front-line staff, support for the integration of diverse cultural groups and provision for special needs. We termed this multi-faceted, localised method of estate rescue a patchwork model for two reasons. First, the many small-scale inputs created an overall pattern of intervention that could be maintained over the longer term because of its ability to generate greater rent income and control costs such as vandal damage. Second, the many elements of the programmes could be fitted together and joined to create an overall impact. It was not a random and fragmented approach; rather it was a locally focused, integrated, flexible, adaptive approach. In these respects the patchwork approach was radically different from the rigid, monolithic design of the estates. It helped to make the estates more akin to successful urban neighbourhoods with their varied and broken up patterns of use but their overall coherence as useable urban assets (Jacobs 1970).

There was not a single solution or a 'big bang' impact from the rescue programmes comparable to the estates' original creation. There was no evidence that such a strategy was feasible or fundable except in a small minority of cases, for the cost of demolition, rebuilding or total restructuring would outstrip the funding abilities of governments for a majority of estate communities. Instead there was a complex set of small-scale, local remedies within the twenty-two measures shown in Table 12.5, which were adapted to each estate community. The overwhelming majority of these actions had a positive effect on conditions. On average eighteen major changes were found in each estate, underlining the diverse range of reforms that were introduced through the rescue initiatives. We estimated that over 90 per cent of the measures on the Continent had a positive impact. This was helped by their strong semi-private management system. Around 75 per cent has a similarly positive impact in Britain and Ireland (see Table 12.6). The weaker

Table 12.6 Impact of remedial measures on overall estate viability

	No. of estates
Average number of remedial measures applied per estate from total of 22 measures	18
Per cent of remedial measures with a positive impact on conditions:	
on the Continent	90%
in Britain and Ireland	75%
Overall impact of remedial measures on estate viability[a]	
Positive	7
Mixed	13
Number of estates that regained viability	16

Source: Author's visits and government research.

Note
a According to government research and landlords, the majority of measures had a positive impact on the particular problem they were addressing, but overall the impact on estates was mixed in a majority of cases.

public management structures, more bureaucratic style and ethos, more limited caretaking role and less clear-cut financial discipline diluted the power and impact of the localised initiatives, although progress was still significant. In all countries landlords believed that a majority of the measures adopted made a significant difference to ground-level conditions, resident satisfaction and financial viability. This was borne out by independent surveys of tenant views in three countries (Gardener 1993; Gifford 1986; SBI 1993).

Social control and the friction principle

While most elements of the rescue initiatives had a positive impact and conditions improved visibly, the overall effect of the programmes was mixed in thirteen out of twenty cases, although most estates regained viability. Table 12.6 summarises the impact of programmes. This limited outcome derived directly from the role of the estates in housing low income and vulnerable households leading to continuing turnover, albeit at a lower level. Table 12.4 shows this. Deep-set problems relating to economic change remained unresolved as they were more influenced by societal and even global pressures, than by local circumstances. Trends in work, in family and in migration were the most important of these.

The increase in control that resulted from the rescue programmes made the estates lettable again, primarily to members of groups that were there already: people in need and people with a weak foothold on the housing or job ladder. This meant that the concentrations of ethnic minority households were growing in the eleven estates where they were already disproportionately concentrated. Levels of unemployment were also rising in social housing generally but particularly in the most unpopular estates such as those in the survey. There was no question of the gentrification of mass estates, although in a limited sense, more stable social conditions were achieved. Excluding disruptive households did not mean excluding people in need (City of Cologne 1995).

The rationale for making estates more socially varied was economic as well as social. It would lead to more and better services, such as shops, due to increased purchasing power, more varied activities, more engagement with the wider community, more people able to play organisational roles locally, more diverse role models, more support for social institutions. The difficulty in making most estates significantly more socially mixed was widely considered a core problem that required continuous compensatory action. In one important sense, however, populations had changed and stabilised, in that tighter management control prevented abuse, enabled enforcement of basic standards and excluded people whose behaviour had been so socially disruptive that conditions for whole communities had been damaged in a way that jeopardised the safety and even the survival of estates.

In France, Germany, Denmark, Britain and Ireland, it was openly accepted by governments and landlords, with strong support from residents, that preventing criminal activities such as drug dealing, violence, intimidation, damage to

235

property and theft was essential if fragile communities were to be strengthened. Only a tiny minority of estate populations was involved in such damaging activities, even at the point where their conditions were most precarious, but a core part of the rescue had involved operating the legal system more equitably in favour of poor communities and reinstating social controls.

One of the ironies of the chaotic decline of mass estates was that while known criminals flouted the law, using fragile communities as cover, an innocent and vulnerable majority was forced to live in conditions of intimidation, harassment and victimisation that the average citizen would consider intolerable (Dauges 1991).

As the estates improved, so too did the quality of life of residents, while their confidence and their access to information and services increased. Their chances of coping and of improving their situation improved in tandem. One important element in the rescue of marginal estate communities was the fact that tenants became less cut off, not least because city governments took far greater interest in their condition and progress. As tenants progressed and moved on, other similar households might take their place, but there was little evidence that estate communities formed a static 'underclass' of people who had given up. The rescue programmes were crucial in reinforcing these links with the city and normalising conditions for people who had previously been surrounded by uncontrollably negative circumstances. (Ministère de la Ville 1993; Jacquier 1991). We likened the process of rescue through stabilisation and normalisation to the friction principle in physics. According to this law, a static object will only move if significant pressure is applied as it has high friction, whereas a moving object will easily accelerate with a small amount of additional pressure as it has low friction. This reflects the experience of mass estates. The worse their conditions were, and therefore the more unstable their communities, the lower their resistance and the smaller the amounts of additional pressure needed for them to spin out of control. Conversely, the more stable they were, the higher their resistance and the less likely they were to spin out of control. For this reason, rescue worked if it enhanced stability, reducing pressures on vulnerable communities. The one depended on the other (Power 1997).

Welfare and work

Unemployment may be the most serious unresolved issue on estates and in European society more generally. Unemployment has soared to new heights in Germany (12 per cent), France (10 per cent) and Denmark (8 per cent). The rate of unemployment has fallen recently in Britain (to around 7 per cent) from very high levels in the early 1990s, partly through economic restructuring, partly through rapid polarisation in incomes and inequality on the American model of less regulated labour markets, a relatively low-wage economy and reduced social underpinning. However, in all countries, including Ireland, which is undergoing economic growth far in excess of the other countries, the numbers out of work for long periods of time are far above the levels dreamt of when the mass estates were built for an employed industrial workforce. In many of the marginal estates we have described more than 50 per cent of households are workless.

To meet the criteria agreed in the Maastricht Treaty, European governments are struggling to reconcile the goals of more jobs at a sustainable and therefore lower, more competitive cost, with the goal of integrative support for the casualties of growth and change. The estate rescue programmes were more successful in the latter than the former. These new pincer movements within the European Union are not yet following the American path of extreme ghetto poverty, racial hostility and the withdrawal of social underpinning from the most vulnerable people and areas; but the price in higher social protection is under increasing political pressure. Under the new pressures of monetary union, high taxes, high unemployment and high welfare costs no longer appear affordable even though the European tradition of civic cohesion will not be discarded lightly. It is precisely this tradition, common to the five countries discussed here, that ensured continuing inputs into mass estates maintaining their housing use and restoring their viability in spite of increasing social pressures. The crucial question is whether European governments, cities and social institutions, such as non-profit landlords, will retain their commitment to buttressing fragile rescue structures, to underpinning precarious communities, and to managing carefully and locally the large, low income housing estates without which large areas of our cities would degenerate into intense ghetto conditions and semi-abandonment as already experienced in American public housing projects and inner city ghettos (Vergara 1996).

Conclusions

The impact of the widespread and localised rescue initiatives in marginal housing areas across five countries illustrates the extent to which support systems can be tapped, reoriented and re-focused to address new problems. The patchwork approach to the problems of mass estates that addresses physical, organisational, financial and social problems together in small-scale, manageable units of organisation has prevented hundreds of precarious communities from continuing on their downward trajectory, arresting decay and re-stabilising conditions. But there are still unresolved questions over the future of Europe's marginal estates. Can areas housing so many needy people retain the hard-won social stability of the rescue period? Can the financial viability of estates be sustained given their high management requirements and the poverty of their residents? Can Europe retain its still viable city cores and re-link its marginal, peripheral estates to the centre? Can the public belt-tightening avoid the withdrawal of support from the poorest communities while supporting a more flexible and competitive European market in the world race for jobs? Will estates remain on the edge or will they become more central to our future as they house youthful populations in need of work in ageing societies where much work remains to be done? Therefore the final question, posed by politicians throughout Europe, is whether we are willing to pay the price of inclusion or would we rather pay the price of exclusion.

In the rapidly changing epoch which gave birth to mass estates and then allocated them to those with little choice, these are unanswered questions. But it seems clear that the model of liberal democracy espoused by Americans which provides strong opportunities but weak underpinning may not fit with the model of social cohesion so arduously constructed over the past half century or more in Europe. Europe is not as rigid and unadaptable as it is depicted by Dahrendorf (1983), and the flexible response to extreme problems demonstrates this. But nor is it any longer as ready to sweep aside its urban legacy as it was after the war, in the face of growing polarisation (Windsor Workshop 1996). In a crowded and increasingly migrant urban world, a more organic approach to cities may be Europe's trump card.

Acknowledgements

This chapter would not have been written without the collaboration of many European housing organisations and individual experts. In particular I would like to thank European representatives at two LSE workshops on this theme, at Cumberland Lodge, Windsor in 1991 and 1996. I would also like to thank American colleagues, especially William Julius Wilson, whose contributions and encouragement in pursuing these ideas inspired further research and confirmed the belief that alternatives to the American urban model might be possible in Europe. Brian Abel-Smith advised on the study throughout.

Note

1 This chapter is based on ten years of research into urban housing problems in Europe. Detailed information about estates is drawn from a survey of twenty large, unpopular, mass housing projects, four in each of the following countries: France, Germany, Denmark, Britain and Ireland; also from five in-depth case studies of one extreme example in each country. Mass housing describes large-scale, publicly funded, post-war, industrially constructed housing developments involving social landlords. Social landlords have received government subsidies to construct mainly non-profit rented housing targeted to meet urgent social needs. Social landlords may be publicly or privately sponsored. But all are regulated by governments and house people in need in exchange for subsidy. High-rise housing describes modern blocks of flats of over five storeys. Visits to the estates were carried out by the author between 1987 and 1995. Information presented here reflects the situation up to 1995. A full account of the study, *Estates on the Edge: The Social Consequences of Mass Housing in Europe*, is published by Macmillan (May, 1997). Government research from each country on marginal estates is listed in the references.

References

AKB (1988) *4 Drifts-omraader I Tastrupgaard*. Tastrupgaard: AKB.

BRBS (1986) *Der Wohnungsbestand in Grossiedlungen in der Bundesrepublik Deutschland*, Heft Nr01/076. Bonn: BRBS.

BRBS (1988) *Städtebaulicher Bericht, Neubausiedlungen der 60er und 70er Jahre*: Probleme und hosungenswege. Bonn: BRBS.

BRBS (1990) *Stadtebauliche Lösungen fur die Nachbesserung von Grossiedlungen der 50er bis 70er Jahre, Teil A: Stadtbauliche und bauliche Probleme und Massahmen, Teil B: Wohnungswirtschaftliche und social Probleme und Massnahmen*, Bonn: BRBS.

BRBS (1991) *Vitalisierung von Grosssiedlungen* (June). Bonn–Bad Godesberg: BRBS.

Burbidge, M., Wilson, S., Kirby, K. and Curtis, A. (1981) *An Investigation into Difficult to Let Housing*: Vol. 1 *General Findings*; Vol. 2 *Case Studies of Post-War Estates*; Vol. 3 *Case Studies of Pre-War Estates*. London: Department of the Environment.

City of Cologne (1989) *Report on the Problems of Kolnberg*. Cologne: City of Cologne.

City of Cologne (1995) *Kolnberg*. Cologne: City of Cologne.

Conrad, C. (1992) 'Réhabiliter ou détruire les grandes ensembles?', *Le Monde* 17 June.

Dahrendorf, R. (1983) *The Limits of Equality*, Stuttgart: Enke.

Dauges, Y. (1991) '*Riots and Rising Expectations in Urban Europe*', LSE Housing Annual Lecture, March, trans. Anne Power. London: LSE Housing.

DoE (Department of the Environment) (1981) *Report on Priority Estates Project visits to Liverpool and Glasgow*, by author (unpublished).

DoE (1995) *Estates Renewal Challenge Fund: Bidding Guidance*, London: DoE.

DoE (Department of the Environment) Ireland (1997) Press Release on the future of Ballymun. Dublin.

Downes, D. (1991) *The Delinquent Solution: A Study in sub-cultural theory*, London: Routledge and Keegan Paul.

Dubedout, H. (1983) *Ensemble Refaire la Ville – Rapport au Premier Ministre du Président de la Commission nationale pour le développement social des quartiers*. Paris: Documentation Française.

Dunleavy, P. (1981) *The Politics of Mass Housing in Britain*. Oxford: Clarendon Press.

Elie, C., Soubeyran, P. and Blery, J.P. (1989) *Roman d'un ZUP*. Villeurbanne: Préfecture Région Rhône-Alpes, Commission Régionale des Quartiers, Association Régionale des Organisms d'HLM de Rhône-Alpes.

Fulrich, M. and Meuter, H. (1987) 'The Legacy of Community House-buildings' in M. Bulos and S. Walker (eds) *The Legacy and Opportunity for High Rise Housing in Europe: The Management of Innovation*. South Bank Polytechnic, London: Housing Studies Group.

Gardener, C. (1993) *An Evaluation of the Ballymun Refurbishment – Report and Appendices*. Dublin: Dublin Corporation.

Gibbins, O. (1988) *Grosssiedlungen, Bestandspflege und Weiterentwicklung*. Munchen: Callwey.

Giddens, A. (1993) *The Consequences of Modernity*. Cambridge: Polity Press.

Gifford, Lord (Chairman) (1986) *The Broadwater Farm Inquiry Report: Report of the Independent Inquiry into Disturbances of October 1985 at the Broadwater Farm*

Estate, Tottenham, chaired by Lord Gifford QC. London: London Borough of Haringey.

Hillebrand, H. (1988) *Wissenwertes uber die Wohnanlage Kolnberg in Koln-Meschenich.* Report to City (24 Feb.). Cologne: City of Cologne.

Holmans, A. (1995) Senior Research Fellow, University of Cambridge. Unpublished commentary on racial issues in social housing, London.

Jacobs, J. (1970) *The Death and Life of Great American Cities.* London: Jonathan Cape.

Jacquier, C. (1991) *Voyage dans dix quartiers Européens en crise.* Paris: L'Harmattan.

Le Corbusier, C.E. Jeanneret (1946) *Towards a New Architecture.* London: Architectural Press.

Levy, F. (1989) *Bilan/Perspectives des contrasts de plan de dévelopement social des quartiers.* La Documentation Française, Paris: Commission Nationale pour le Development Sociale des Quartiers.

McGuire, C.C. (1981) *International Housing Policies: A Comparative Analysis.* Lexington NJ: Lexington Books.

Ministère de la Ville (1993) *Politique de la Ville,* Paris: Ministère de la Ville.

Mitterrand, F. (1993) *Introduction, Politique de la Ville,* Paris: 1993

North Islington Housing Rights Project (1976) *Street by Street: Improvement and Tenant Control in Islington.* London: Shelter.

OECD (1987) *Urban Housing Finance.* Paris: OECD/Albin Michel.

Peach, C. (1995) *Ethnicity in the 1991 Census.* Office of National Statistics, London: HMSO.

Power, A.E. (1993) *Hovels to High Rise: State Housing in Europe since 1850.* London and New York: Routledge.

Power, A.E. (1996) *Perspectives on Europe.* London: Housing Corporation.

Power, A.E. (1997) *Estates on the Edge: The Social Consequences of Mass Housing in Europe.* London: MacMillan.

Power, A.E. and Tunstall, R. (1995) *Swimming Against the Tide: Polarisation and Progress on Twenty Unpopular Council Estates.* York: Joseph Rowntree Foundation.

Power, A.E. and Tunstall, R. (1997) *Dangerous Disorder.* York: Joseph Rowntree Foundation.

Provan, B. (1993) *A comparative study of French and UK government programmes to tackle the physical management and social problems of post-war housing estates,* unpublished PhD thesis, University of London.

Quilliot, R. and Guerrand, R H (1989) *Cent Ans d'Habitat: une utopie realiste.* Paris: Albin Michel.

Reich, R. (1993) *The Work of Nations: Preparing Ourselves for 21st Century Capitalism.* New York: Vintage Books.

Salicath, N. (1987) *Danish Social Housing Corporations, Vols I and II.* Co-operative Building Industries Ltd, with the support of Danish Boligministieret, Copenhagen.

SBI (1993) *Bedre Bebyggelser – bedre liv?* (Better Housing Estates – Quality of Life?) Results from an evaluation project by the Danish Building Research Institute, Town Planning Report 65 Horsholm: State Building Research Institute, Danish Ministry of Housing.

SCIC Gestion (1995) *Newsletter and visit report.* Paris: SCIC.

Vergara, J. (1996) *The New American Ghetto.* New Brunswick: Rutgers University Press.

Vestergaard, H. (1993) *Improvement of Problematic Housing Estates in Denmark – An Evaluation of Results*. Horsholm: Danish Building Research Institute.

Wilson, W.J. (1996) *When Work Disappears*. New York: Alfred Knopf.

Windsor Workshop (1991) LSE European Social Housing Workshop, Cumberland Lodge, Windsor, 10–12 April.

Windsor Workshop (1996) LSE European Social Housing Workshop, Cumberland Lodge, Windsor, 25–27 June.

13

WESTERN EUROPEAN HOUSING POLICIES

Convergence or collapse?

Mark Kleinman

Introduction: the bifurcation of housing policy[1]

Over the last twenty years, housing policy has become an ambiguous term. Increasingly to talk about housing policy means to talk about two very different sets of concerns, issues and possible solutions. One set of concerns relates to the circumstances of the majority, who are mainly well housed and can reasonably expect to be better housed in the future. Although these households are mainly in the market rather than the social sector, their housing outcomes are strongly affected by various types of public intervention, those which ensure continuity and reasonable market conditions in the private sector. Ensuring this means that government assumes responsibility for providing a legal framework for the enforcement of contracts; for maintaining the supply of finance; for providing output to some degree, especially counter-cyclically; for defining a land use planning framework; for maintaining affordability through subsidies, especially to owner-occupiers; and, perhaps most importantly, for ensuring steady economic growth and a reasonably high level of employment.

The other set of concerns relates to the circumstances of the disadvantaged, who are badly housed or homeless, whose prospects of future betterment are uncertain, and whose residential segregation, in many cases, compounds social and economic inequality. Moreover, if (as seems likely) economic integration in Europe will lead both to a greater average standard of living and to wider disparities between individual households and between different geographical areas, then there will be a strengthening of such trends.

This second aspect to housing policy, which involves issues such as homelessness, the management of social housing, the availability of means-tested housing allowances and so on, relates to a minority of the population, a minority which is increasingly segregated or at least differentiated from the majority geographically, ethnically or in terms of household type. Whatever the formal appearance, such policies and their associated expenditures are consented to by the majority, not as a type of collective provision, but as a form of altruism, (helping the poor); or as an insurance payment against riot, theft or social disorder; or as socially necessary expenditure (because low-paid but essential workers need to live somewhere). 'Housing policy', as defined in this narrow way is thus mainly concerned with social housing (including its privatisation). As such, it may seem to have little direct influence on the interests of the majority of the population in most Western European countries who do not live in social housing.

The convergence debate

The focus of many comparative housing policy studies is primarily empirical, with theory kept to a minimum. For example, in one of the earliest comparative studies, McGuire (1981) identified four policy 'cycles': first, acute shortage and an emphasis on stimulating production; second, an objective of increasing the size of units; third, a focus on higher quality; and fourth, reduction in the state burden by switching from indiscriminate to targeted subsidies. The concept of policy 'cycles' is not a particularly useful one. By suggesting a process of circulation and repetition it entirely misses the idea of a *dynamic* change from one phase of policy to the next. It suggests a 'natural' progression from one set of problems to the next, ignoring the complex interplay between structural change, political action and policy change.

A decade later Boelhouwer and van der Heijden (1992) in their seven-country study identify four policy 'stages' in the period 1970–1990. These are: a quantitative stage; a qualitative stage; an emphasis on distributive issues; and finally the re-emergence of housing shortages. Boelhouwer and van der Heijden's typology is more flexible than McGuire's; stages do not necessarily coincide for all countries and some stages can occur simultaneously in the same country.

Boelhouwer and van der Heijden's typology is more useful as a descriptive tool, but does not take us very far in terms of providing an explanation as to why successive stages of policy should occur at all, let alone occur at different times in different countries. They do however correctly emphasise the (often implicit) role that convergence theories have played in comparative housing research: 'Despite . . . criticisms, convergence theory forms the theoretical framework of much social scientific research, including most (comparative) housing research' (Boelhouwer and van der Heijden 1992: 5). These authors follow Schmidt (1989) in tracing this influence back to the work of Donnison (1967). The core idea is that similarity of economic and demographic developments in different countries, will mean that housing policies will converge, despite differences in politics, ideology and institutional arrangements.

This approach to the comparative study of housing policies can thus be seen as one example of what Wilensky *et al.* (1987) identified as the dominant approach to welfare state research, i.e. the explanation of rising welfare state expenditures as the consequence of economic development and socioeconomic convergence across countries. This strand of research is close to what Esping-Andersen calls the 'systems/structuralist' approach: industrialisation makes social policy both necessary – because pre-industrial modes of social organisation are destroyed – and possible – through the rise of modern bureaucracy. Hence, 'this approach is inclined to emphasize cross-national similarities rather than differences; being industrialized or capitalist over-determines cultural variations or differences in power relations' (Esping-Andersen 1990: 12).

Many comparative studies of housing policies in Europe make use of the concept of policy convergence. A good example is the work of Ghékiere (1991, 1992). Ghékiere argues for 'certain types' of convergence in the housing policies of the EC 12 in the 1980s; specifically, that there is a 'convergence model', relevant to a growing number of countries which transcends the national context, and even the political colour of the government in power (Ghékiere 1992: 205). He argues that though there are important differences, particularly between the countries where there has been a sharp break with the previous model – a group in which he includes the UK, Spain, Ireland and the Netherlands – and those where there has been more continuity and some continuing adherence to the principle of social economy (especially France and Germany), the key change is from the 'Long Boom' model of massive state intervention, through state building programmes and rent control, and the post-1975 model.

He defines this latter convergence model as being ideologically driven by the doctrine of the minimal state, imported from the USA, welcomed and developed in the 'Anglo-Saxon' countries (presumably the UK) but progressively adopted in other European countries, and 'even' within the EU organisations. The model comprises disengagement of the state from housing, abolition or liberalisation of rent control, and the determination of investment levels by private rather than public activity. The Anglo-Saxon model is posited on the hypothesis that because housing need and supply are globally in balance, disequilibria (that is, unmet needs) persist only because of rigidities in the market, not because of inadequate supply. Hence the main regulatory role should fall to the market, with the state's role being limited to the correction of dysfunctions. The model typically is implemented through a shift to targeted support, decentralisation of housing policy and the dismantling of specific systems of actors and circuits of financing ('banalisation').

Ghékiere admits that this model is 'rather caricatured' (1992: 219) and cannot be strictly applied to each of the (then) twelve member states of the EC, but argues that the convergence model guides the direction of policy. Its application '*pure et dure*' is restricted to the UK, Ireland and the Netherlands, countries with 'liberal' or 'ultra-liberal' governments. Policy in Germany, Denmark and France, by contrast is characterised by the 'continual adaptation' of the instruments of intervention, so there has not been such a break with previous policies. That is, Ghékiere argues that these countries have more flexible housing policy mechanisms, as opposed to the more rigid and hence antiquated systems elsewhere.

This is an example of what we can term the 'weak form' of convergence theory. In its strong form, convergence theory insists that social and economic convergence across countries follows inevitably from the 'logic of industrialism'. In its weak form, the argument is simply descriptive: policies happen to be converging.

How useful or accurate is the notion of policy convergence? Even in its weak form, the convergence case is unconvincing. Ghékiere draws too great a distinction between Anglo-Saxon radicals and Rhineland pragmatists. Housing policy in Britain after 1979 had many continuities with earlier periods – increased control by central government, bipartisan support for owner-occupation, the shift towards means-testing, mechanisms to raise social and private rents, etc. Ghékiere also claims that in Britain, social rented housing was financed by 'direct state loans, rarely repaid', and hence the costs of the system rapidly became prohibitive, while in Germany and France, the system of public loans was progressively diversified. In fact, local authority loans in Britain have always been raised directly or indirectly (via the Public Works Loans Board) from the market, and do of course get repaid over time, the loan charges falling on the Housing Revenue Account. Indeed, it is precisely the low debt encumbrance of many smaller British local authorities as these loans were paid off that made stock transfers attractive as a form of privatisation from the mid-1980s on. At the same time, as shown in Lefebvre *et al.* (1991) among others, the French system of financing social housing has encountered considerable difficulties caused by disintermediation reducing the balances in the Livret A accounts. Similarly, as Harloe (1995) has pointed out, it is 'corporatist' Germany that, together with Britain, has gone furthest in privatising its social rented stock.

Ghékiere claims that twelve years of ultra-liberal policy in Britain have 'degraded' the UK situation and multiplied homelessness and exclusion through reduced supply. While the trend in British housing policy is not disputed, it is important to point out that even by the mid-1990s, Britain had a much larger social sector than either Germany or France, with a legal right of access to housing for defined homeless households. Homelessness may appear to be a greater problem in Britain than in France or Germany precisely because of the existence of a legally enforceable right to housing for some groups. Statistics on priority homelessness in Britain measure a channel of access into social rented housing rather than the level of social exclusion. Of course this is not to deny that homelessness, including street homelessness, is a problem in Britain, nor to deny that it has got worse over the last twenty years. It is simply to point out that the data for accurate inter-country comparisons do not yet exist (Bayley 1994).

Furthermore it has been in France rather than in Britain that 'Anglo-Saxon' monetary orthodoxy in the management of the economy was applied most consistently through the 1980s and 1990s. Britain, despite the hard-line rhetoric, has swung wildly between hairshirt 'sado-monetarism' and wild binges of debased Keynesianism in the form of tax cuts and 'dashes for growth'. Clearly, there *are* important differences between Britain on the one hand and France and

Germany on the other – in terms of ideology, institutions and practice – but these differences require a more sophisticated formulation to capture them.

More generally, the convergence debate does need to be approached carefully. Apparent similarities do not imply convergence. At the heart of the debate is a process by which various social, economic and political trends, common to more than one country, are mediated through a set of national institutions, policy traditions, history and culture which are unique to individual countries. The result of this is a complex pattern which requires careful analysis.

It is the apparent similarities between housing policies in different countries – e.g. the supposed disengagement of the state, the switch from producer to consumer subsidies, etc. – which lead commentators such as Ghékiere to speak of convergence in European housing policies. But the term convergence – at least in its strong form – entails more than mere similarity. It conveys the idea of a causal process by which the housing systems of different countries *necessarily* become more alike over time. Indeed the concept of convergence is grounded in the sociological literature on the theory of industrialism; it implies the view that industrialisation is the motor of social change, and that all industrial societies will converge towards a similar form of mixed economy with substantial state intervention (Harloe and Martens 1983). Convergence theory implied that the logic of industrialism would lead not to class conflict, but to elite leadership and mass response. Pluralistic industrialism would generate convergence, as there were in effect no ideological choices left. But in practice, the pattern has been one of divergence, rather than convergence (Goldthorpe 1984).

Harloe (1995) in his major study of social housing in six countries (Britain, France, Germany, the Netherlands, Denmark and the USA) argues that in comparative studies of housing policy, one must account for both similarities and differences across countries. Harloe quotes Gourevitch with approval:

> Each country is affected by these twin factors: the force of epochs, which cuts across the particularities of circumstance, and the force of national trajectories, which expresses the features specific to each nation's history.
>
> (Gourevitch 1986: 217)

Comparative housing studies must therefore accommodate both the common patterns of social and economic change and also specific national circumstances.

Although strict typologies of welfare state regimes tend to break down when confronted with the complexity and hybridity of actually existing systems, the broad distinction between the Anglo-Saxon British welfare state and the corporatist systems of France and Germany has been found to be meaningful (Kleinman 1996). Britain, in comparison with France and Germany has a greater commitment to owner-occupation; a greater concern with housing policy as a redistributive tool rather than a reflection of the status order; far greater centralisation of policy with a minor role for sub-national government in policy formulation;

greater mistrust of the state as an active agency in implementing change. If we ask Esping-Andersen's question, posed with regard to social insurance, the choice of 'whether to allow the market or the state to furnish adequacy and satisfy middle-class aspirations' was answered, for housing policy, with a clear decision in favour of the market.

As the comparative study of housing policies and housing systems has mush-roomed over the last ten to fifteen years, it is the peculiarities of the English which often come to the fore (sometimes literally the English: Scotland has sev-eral points of greater congruity with continental housing systems, such as a trad-ition of apartment-dwelling in large cities). These include the obsession with homeownership; the weak private rented sector; the deregulated and *laissez-faire* environment, the rejection of national planning and output targets; the domin-ance of local authorities in the rented sector; and particularly, the pauperisation and means-tested dependence of the social sector.

How can we reconcile both the similarities and differences between Anglo-Saxon and corporatist policy regimes? The convergence hypothesis can take one of two opposing forms. In the first version, Europe is seen as converging to the Anglo-Saxon model. Britain is Airstrip One, the unsinkable aircraft carrier to which American free market ideologies are flown for re-export to regional mar-kets. As with Japanese cars and Korean hi-fis, Britain is the point of entry through which foreign state-of-the-art ideology gains access to previously sheltered European markets. As free-market liberal ideology becomes ever stronger, both corporatist welfare states and even the saintly European Union succumb.

In the alternative version, the process is reversed. Britain is an aberration – an atavistic nation-state of a kind which has become obsolescent elsewhere in West-ern Europe, and which has been kept alive only through the continuing effects of the powerful and unique political phenomenon known as Thatcherism. How-ever, Britain's institutions desperately require modernising, and this process will inevitably take Britain closer to the norm of the modern European welfare state, which is essentially the traditional Rhineland corporatist state – somewhat modi-fied and more market-friendly, but clearly recognisable through its emphasis on consensus, modernisation and partnership.

The evidence that I present in my book shows that neither of these versions is accurate (Kleinman 1996). First, it is by no means clear that we can define an 'Anglo-Saxon' housing model. As Harloe shows, housing has always been an anomalous and ambiguous component of the welfare state. The majority of cit-izens in all Western European countries as well as in the USA have been housed by private sector activity, with only a minority in the social sector. Indeed, apart from the Netherlands, the largest social sector in Europe is found not in any of the corporatist or social-democratic countries, but in Britain. While the promo-tion of owner-occupation and the deregulation of finance markets have pro-ceeded much further in Britain than elsewhere, the differences seem to be those of degree, not those of kind. If there was a distinct Anglo-Saxon model we would expect categorical, in-kind differences.

Second, certain aspects of the French and German housing systems reflect a more thorough-going liberal ideology than in Britain. The course of French housing policy in the last decade has been strongly influenced by the strict adherence of the French state to the principles of monetary and fiscal orthodoxy. These principles are often closely associated with the British state under Thatcher and Major, but in practice were often honoured more in the breach than in the observance. Similarly it was in Germany, not in Britain, that the entire social rented stock was in effect privatised in the mid-1980s, and a consideration of German housing policy more generally shows a close concern with market processes and a clear vision of state activity supporting and not replacing market forces.

So convergence theory, in both its strong and its weak form, is rejected as an overarching explanation. Nevertheless, we can identify important common *themes* in policy development in the recent period. These include:

- a greater role for markets in the production, allocation and financing of housing
- the promotion and encouragement of owner-occupation
- a switch from new building to renovation
- an emphasis on the cost-effectiveness of policy measures
- greater market orientation and private sector ethos of social housing agencies
- at the same time wider responsibilities of social housing agencies to house the poor and cater for a range of disadvantaged and special needs groups
- targeting of social housing subsidies as part of a widespread trend in social policy towards selectivity
- deregulation of housing finance markets
- an emphasis on 'holistic' solutions to problems of poor housing and urban regeneration

Having rejected policy convergence as an overarching explanatory concept, what can we put in its place? The complexities of the interaction between system change and policy change, and between common economic and social forces and country-specific institutions and practices make it unrealistic to propose a simple alternative. But rather than policy convergence, the idea of policy *collapse* appears to be a more accurate way of trying to capture twenty years of change in housing policy in the three countries.

The collapse of housing policy

As traditionally understood, housing policy was about estimating housing needs, setting quantitative output targets, boosting house-building (especially in recessions), raising the average physical condition of the stock, removing substandard housing, and pursuing a goal of a decent home for every household.[2] It was these sorts of considerations that dominated discussions of housing policy in the three decades after 1945, and indeed the debate on the 'Housing Question' for a half-century or more prior to that.

The ending of the crude shortage of housing units in many European countries led to a weakened perception of housing as a national issue. Housing began to be seen more as a range of local, sectional and special needs issues. Policy has shifted from mass solutions to individual solutions. This shift was more pronounced with regard to housing than with regard to other social policies such as health or education, both because of the nature of housing, with more of the characteristics of a private good (Barr 1987: 408–410) and also because of the fact that housing was never provided in Western Europe as a universal social service.

At the same time, the state has tried to reduce its own share of housing costs, and thereby to bring about an increase in the share borne by households themselves. Several factors are relevant here. With economic growth, real incomes rise and households can bear a larger proportion of costs without hardship. At the same time, central and local governments facing 'fiscal crisis' through the continual growth in assumed responsibilities and hence expenditures, have sought to pull back from commitments. But rising average prosperity has been accompanied by falling real incomes for a minority of the population, and hence by increases in both poverty and inequality. Although the state has attempted to reduce its share of housing costs, in practice this has been difficult to implement, as cuts in producer subsidies lead to greater take-up of means-tested allowances, and economic restructuring raises unemployment and hence benefit payments.

Alongside this, there has been a trend towards devolution and decentralisation. Again, this is a complex process. It is clear that central governments have become more aware of the boundaries of their own competence and abilities. But the actions of central government in redefining a narrower role for itself have both negative and positive consequences for local areas. Decentralisation can be a means of encouraging pluralism and sharing power, or it can simply be a way of offloading responsibilities by devolving responsibility but not power nor resources (Johnson 1987).

Housing policy in the traditional sense has virtually collapsed. It has bifurcated, that is, split apart, leaving behind two separate but related sets of policy issues. On the one hand there is a set of issues which relate to concentrations of poverty, associated with economic restructuring and social disintegration. These are not fundamentally bricks and mortar issues, nor even about housing management and housing finance. They are increasingly about social dysfunction, about the collapse of communities, about the impacts of mass unemployment and poverty on everyday life.

At the same time, we have a second set of issues which are far removed from this grim picture, but equally distant from the earlier concerns of output targets and physical standards. These issues revolve mainly around the continuing expansion of owner-occupation as a visible sign of economic and social success, both for the individual household and for society as a whole, including maintaining the value of the asset to the households which have purchased it. The key development is the normalisation of property ownership as a route to social stability.

Despite the rhetoric about the fight against social exclusion, the reality is that the European political economy is now founded in practice on the acceptance at a more or less permanent level, of a continuing divide between the haves and have-nots in each country. In housing policy, this underlying belief finds expression in the retreat of national governments from responsibility for achieving more equal housing outcomes. As the divide grows, policy bifurcates between, on the one hand, measures to maintain market stability for the majority, either in terms of mass owner-occupation or a more balanced private renting/owner-occupation split, and, on the other hand, measures to alleviate some of the worst excesses for the poor, while transferring responsibility from national to local, or even community level. The rhetoric of so-called holistic solutions supposedly integrating housing, employment and welfare aspects has been accompanied by the state withdrawing from its responsibility for full employment and reducing its welfare commitment. The emphasis on empowerment, localism and bootstraps approaches needs to be seen in this light.

Housing policy as welfare division

In the collapse of housing policies since the mid-1970s we can read part at least of the story of how both economy and society have changed since the end of the long boom. We see in housing the reflection of the broader pattern of welfare division and polarisation.

As a French study puts it, 'more markets mean more exclusion' (Lefebvre *et al.* 1991). This is not unique to housing, but we see it more clearly in the housing sector than in other aspects of the welfare state. This is for two reasons. First, because housing has always been the 'wobbly pillar of the welfare state', i.e. a mix of public and private provision, and never quite sure if it was part of the welfare state or not. Second, because, as such, housing is as much or more affected by economic as by social policy. Through studying the collapse of housing policy over the last twenty years in Britain, France and Germany, and through understanding how little effect the European dimension is going to have on the fundamentals of policy, we can also learn something about what has happened to welfare over that period, and how our view of what policy can achieve has changed over that time.

Clearly, the study of housing policy in Britain, France and Germany reveals a complex pattern which cannot be reduced to a simple formula. There are economic, political and social forces which are common to all three countries (as well as to many others), but these forces are mediated through very different institutions and political structures. Moreover the effects of these institutions and structures is more than just to speed up or slow down the pace of change and adjustment – they leave a profound impression on both policy and its implementation. It is here that the concept of path dependence can be useful – countries become locked into particular patterns of policy development at an early stage, for reasons that may be historical, deliberately chosen, or the product of accident. Once locked in, this pattern then constrains future development.

All explanations of comparative policy development therefore involve some combination of common structural forces on the one hand, and different institutional and political mechanisms on the other. So, while Esping-Andersen, for example, emphasises the latter, Harloe puts more weight on the former. Ultimately, the exact proportions in the mixture are perhaps only a matter of taste; I am not concerned here to prolong further the debate on structure and agency, but simply to reaffirm that any adequate causal explanation requires the presence of both components.

Whatever the relative contributions of structure and agency which best explain the current state of housing policy in Britain, France and Germany, the general *direction* of policy is clear, in all three of the countries studied. Housing policy, in the sense in which the term was understood at the zenith of the post-war welfare state, has collapsed or is collapsing, leaving behind a bifurcated or polarised set of policies towards housing which both reflect the division of modern European societies into a relatively contented majority and an impoverished minority, and by and large promote the acceptance of this state of affairs.

Underlying causes

Contrary to much wishful thinking on the left and centre-left, the European dimension does not offer an alternative vision to this, but largely reflects and reinforces what has become the consensus among policy makers: that primacy must be given to the free play of market forces; that government is mainly the problem not the solution; that little can be done about growing economic inequality; that mass unemployment and poverty are probably an inevitable part of the political and economic landscape.

Furthermore, there seems to be relatively little prospect of any radical change to this position. Not only do the vast majority of politicians and senior officials subscribe to this limited and pessimistic vision of government's role, but it is a view increasingly shared by electorates also.

As Paul Krugman has argued, the key economic variables which affect people's standard of living are productivity, income distribution and unemployment. Other variables such as inflation and the budget deficit have only indirect effects (Krugman 1990). Moreover, of these productivity is the most important:

> Productivity isn't everything, but in the long run it is almost everything.
> A country's ability to improve its standard of living over time depends
> almost entirely on its ability to raise its output per worker.
>
> (Krugman 1990: 9)

For reasons that are unclear, despite being much debated, productivity growth in the USA and Western Europe slowed after 1973. The ending of the Long Boom continues to define the era in which we are living – in Krugman's phrase, the age of diminished expectations. He is referring to the expectations of individual households about their future levels of economic welfare, but the phrase can equally well be applied to collective expectations about the ability of welfare states to provide full employment, social protection, and high standards of public and private welfare. The stuttering of the great Keynesian economic machines has thrown into jeopardy not just the post-war consensus on collective management of the economy, but also the continuous growth in the welfare state which was predicated on the success of such management.

More specifically, in Europe there has been the inexorable rise and persistence of unemployment. Unemployment in the EU 12 was not only far lower in the early 1970s than it is now (below 4 per cent), but at that time was also below the level of unemployment in the USA (although still above Japanese levels). Unemployment in Europe rose steadily throughout the 1970s, and in the 1980s and 1990s has remained considerably above levels in the United States, and far above those in Japan. Explanations abound for this phenomenon. For free-marketeers the culprit is Europe's regulated labour markets, compared with the more liberalised US system. As evidence, they point to the much lower rates of job creation in Europe, and the consequent lower employment rates. That is, Europe's growth in unemployment has largely been caused not by demographic factors (increases in the labour force) so much as from an inability to create enough jobs. For interventionists, the reasons are more directly to do with government policy (Nolan 1994). A decade or more of deflationary policies, in which the control of inflation was prioritised over full employment, has left Europe's economies working at well below capacity, and with a consequently large volume of demand-deficient unemployment (see Michie and Wilkinson 1994). The solution is not yet more market liberalisation, but rather active policies of demand management to boost output and employment. These policies now look even further away with the institutionalisation of deflationary policies in the mechanisms of the ERM and the Maastricht Treaty. Bean (1992) argues that European unemployment is high for a combination of both classical and Keynesian reasons. But depressingly, he concludes that we are not much further on than ten years ago in understanding why this should be.

These economic developments have put similar pressures on social policies in the three countries in two ways. First, the slowdown in economic growth, coinciding with large, sometimes explosive growth in welfare expenditures has raised the issue of 'can the welfare state be afforded?' to the top of the political agenda. As Rose (1986) pointed out, it is not the growth in public expenditure *per se* which is crucial, but rather the 'front-end load', that is, the proportion of economic growth absorbed by public expenditure. In the seven OECD countries studied by Rose, this increased from 28 per cent (1952–1960) to 47 per cent (1961–1972) to 147 per cent (1973–1982).

Second, the increase in unemployment, together with other factors such as the growth in the numbers of single-parent families and the widening distribution of income, have led to increased numbers of poor people, and consequently greater demands on social protection systems and other forms of social welfare. To this might be added changes in the demographic structure, with increased numbers of elderly relative to working-age population, although many of the more apocalyptic claims about the consequences of an ageing population can be discounted.

Together, then, these economic developments create similar pressures on social policies in our three countries and elsewhere. Moreover, the concentration of problems amongst a minority of the population, and the growing gulf between those whose living standards remain tolerable or better, and those whose real position worsens, will by its very nature weaken support for universal welfare programmes. Hence there is a strong interconnection between these economic factors – changes in the material basis of households' lives – and more directly political or ideological factors.

In addition, there has been a shift in political attitudes among policy makers and politicians, and possibly also among voters. It would be difficult as well as time-consuming to tease out the complex relationships between structural economic changes and changes in ideas and political values in the 1970s and 1980s. But it is clear that there was a widespread shift to the right, away from planning and towards the market, away from government intervention and towards market liberalisation.

A secondary question is the degree to which this change in core economic beliefs and social philosophy is simply an elite phenomenon, or whether it is shared by the majority of the population also. Radical right governments in Britain were elected on minority shares of the vote, the Conservatives achieving a remarkably consistent 41–42 per cent in the elections of 1979, 1983, 1987 and 1992. France was governed by socialist governments for much of this period, while Germany's federal government remained a coalition which represented social-Catholic as well as neo-liberal elements (Mangen 1991).

Furthermore, in Britain at least, there is considerable survey evidence that the welfare state continued to be extremely popular with voters in the 1980s, and that voters expressed willingness to pay higher taxes to support the welfare state (see Taylor-Gooby 1991: Ch. 5 for a discussion). In 1983, 32 per cent of the British population supported increased taxes and spending on health, education and social benefits, compared with only 9 per cent who wished both taxes and spending to be reduced. By 1989 the figures were 56 per cent and 3 per cent. Nevertheless, voters continued to return governments that favoured tax cuts over increased spending. The explanation for this paradox may lie in the relatively low priority voters give to the welfare issue as a determinant of how they vote, or it may simply be evidence of the gap between voters' theoretical beliefs and their practical concerns.

Whatever the degree of genuine ideological change among either elites or masses, and whatever the strengths or shortcomings in the new market ideology, the essential point for our argument here is that at the very least, these political developments in the last two decades represent a considerable loss of faith in the post-war Keynesian social-democratic project. The reduction in economic momentum, the chronic problem of inflation and the re-emergence of mass unemployment, above all the seeming inability of traditional policy mechanisms – or indeed *any* policy mechanism – to have much impact on these problems have led to citizens' having much weaker beliefs in government's ability to do very much to ensure economic growth and improve social conditions. This represents a widespread and significant development in the political economy of the Western European countries. Of course there are significant differences in how these issues were played out in Britain, France and Germany, and the actual institutional and policy environments had significant effects in terms of how severe the consequences were and who bore them. But the common experiences of disillusionment should not be ignored. This is the age of diminished expectations not just about individual economic advancement but also about the effectiveness of government.

So while there will continue to be debate among elites, between political parties, and among the public too, about the *specifics* of housing policy in each country – where the limit of owner-occupation is, the balance between means-tested and general support, what the most effective ways of targeting diminishing resources are, and so on – there seems little scope for changing the broad limits which have been imposed. Yet underlying this the goals of a truly holistic – that is, universal – housing policy remain the same: a decent home for all, the improvement of the housing stock, a reduction in inequality. But to achieve this would require not just linking certain aspects of housing to other social issues (what is meant by holistic in practice) but where necessary challenging the economic orthodoxy that limits the scope and function of social policy in general. This would mean recognising housing as a social investment, looking at the housing system as whole, prioritising need, promoting fairness, investment and quality rather than tenure-specific goals, linking housing to the debate about equality and polarisation and seeking the solution to current housing problems in their root causes of unemployment, family breakdown and political unresponsiveness. Only in this way might the bifurcation of housing policy be reversed and the two paths brought together again.

Notes

1 This chapter draws on the author's extensive research on housing policies and outcomes in Britain, France and Germany. See Kleinman (1996).
2 The definition of 'household' is itself problematic, of course. In particular, the post-war period has seen a shift in the commonly accepted definition from being essentially synonymous with 'family' to something much broader.

References

Barr, N. (1987) *Economics of the Welfare State*. London: Weidenfeld and Nicolson.

Bayley, R. (1994) 'A Right to Housing, but Not to a House', *Inside Housing* 1 July.

Bean, C. (1992) 'European Unemployment: A Survey', London School of Economics Centre for Economic Performance Discussion Paper 71, March.

Boelhouwer, P. and van der Heijden, H. (1992) *Housing Systems in Europe, Part 1*. Delft: Delft University Press.

Donnison, D. (1967) *The Government of Housing*. Harmondsworth: Penguin.

Esping-Andersen G., (1990) *The Three Worlds of Welfare Capitalism*. Cambridge: Polity Press.

Ghékiere L. (1991) *Marchés et politiques du logement dans la CEE*. Paris: Documentation Française.

Ghékiere L. (1992) *Les Politiques du logement dans l'Europe de demain*. Paris: Documentation Française.

Goldthorpe, J. (1984) 'The End of Convergence: Corporatist and Dualist Tendencies in Modern Western Societies', in J. Goldthorpe (ed.) *Order and Conflict in Contemporary Capitalism*. Oxford: Oxford University Press.

Gourevitch, P. (1986) *Politics in Hard Times: Corporative Responses to International Economic Crisis*. Ithaca and London: Cornell University Press.

Harloe, M. (1995) *The People's Home?* Oxford: Blackwell.

Harloe, M. and Martens, M. (1983) 'Comparative Housing Research', *Journal of Social Policy* 13: 255–277.

Johnson, N. (1987) *The Welfare State in Transition: The Theory and Practice of Welfare Pluralism*. London: Harvester Wheatsheaf.

Kleinman, M.P. (1996) *Housing, Welfare and the State in Europe*. Cheltenham: Edward Elgar.

Krugman, P. (1990) *The Age of Diminished Expectations*. Cambridge, MA: MIT Press.

Lefebvre, B, Mouillart, M. and Occhipinti, S. (1991) *Politique du logement: 50 ans pour un échec*. Paris: L'Harmattan.

Mangen, S. (1991) 'Social Policy, the Radical Right and the German Welfare State', in H. Glennerster and J. Midgley (eds) *The Radical Right and the Welfare State*. Hemel Hempstead: Harvester Wheatsheaf.

McGuire, C. (1981) *International Housing Policies*. Lexington, NJ: Lexington Books.

Michie, J. and Wilkinson, F. (1994) 'The Growth of Unemployment in the 1980s', in J. Michie and J. Grieve Smith (eds) *Unemployment in Europe*. London: Academic Press.

Nolan, P. (1994) 'Labour Market Institutions, Industrial Restructuring and Unemployment in Europe', in J. Michie and J. Grieve Smith *Unemployment in Europe*. London: Academic Press.

Rose, R. (1986) 'Common Goals but Different Role: The State's Contribution to the Welfare Mix', in R. Rose and R. Shiratori (eds) *The Welfare State East and West*. New York: Oxford University Press.

Schmidt, S. (1989) 'Convergence Theory, Labour Movements and Corporatism: The Case of Housing', *Scandinavian Housing & Planning Research 6: 83–101*.

Taylor-Gooby P. (1991) *Social Change, Social Welfare and Social Science*. Hemel Hempstead: Harvester Wheatsheaf.

Wilensky, H.L., Luebbent, H.L., Mahn, G.M. *et al.* (1987) *Comparative Policy Research*. Berlin: Gower.

Part IV

DIVERSITY IN THE
EUROPEAN UNION

14

THE NORDIC COUNTRIES

Sirpa Tulla

All Nordic countries are sparsely settled – Denmark though is more densely populated than the others.[1] Their languages are related – except the Finnish language. Three out of the five – Denmark, Finland and Sweden – belong to the European Union; Norway and Iceland are members in the European Economic Area (EEA). All Nordic countries share, however, social and cultural values best known in connection with the concept of a Nordic welfare state.

The Nordic Council was established in 1953 to promote co-operation between the parliaments and governments of Denmark, Finland, Iceland, Norway and Sweden. It is a purely consultative organisation; there exists no supranational element in its organisational structure. Questions of common interest to the Nordic countries are discussed in the Council and recommendations are made to the governments. Piecemeal harmonisation of legislation on selected policy issues has been the chosen strategy in Nordic integration.

Turning the Nordic countries into one unit with respect to social rights was a major goal already at the very beginning of Nordic integration. A common labour market and a passport union between the Nordic countries were important elements in the integration process. The passport requirement was abolished in 1954 for Nordic citizens travelling within the Nordic countries. The Labour Market Convention on free movement of labour was also signed in 1954. The Social Security Convention signed in 1955 entitled the citizens of one Nordic country, but living in another, to the social benefits of the country they were residing in. The Nordic countries were the first European group of countries to develop a community beyond national borders in these fields (Solem 1977: 113).

There is, however, no evidence on economically significant migration within Nordic countries. The only quantitatively large migration was that from Finland to Sweden in the late 1960s. The number of Nordics as a proportion of total immigration has actually fallen. Both in the Nordic Common Labour Market and in the European Union, labour has reacted little to the opportunity of free movement. Comprehensive social security could be one factor which explains the low intra-Nordic mobility (Fischer and Straubhaar 1996: 178, 209).

Housing authorities have been meeting regularly since the mid-1950s under the umbrella of the Nordic Council. Exchange of information and joint research projects in housing and building are the main focus of these meetings. No integration in housing policies has been attempted; nor has it been desirable. There are striking differences between the Nordic countries in housing provision despite the shared community values. In this chapter some Nordic approaches to housing and integration are presented. The point of view is that of Finland, but comparisons are made with the other Nordic countries.

Housing and the Nordic welfare state

Economic and social change

The core features of the Nordic welfare state include a broad coverage of social security at an adequate level, universal and equal public services for all and extensive income transfer systems. Universality is stressed as a means of preventing the exclusion of the individual and of creating equal opportunities for all. Social security systems in Finland and the other Nordic countries still rely fundamentally on the idea of the Nordic welfare state, although in public discussions the concept itself has sometimes become more like a relic from the heyday of social democratic planning.

The economic recession of the early 1990s hit Finland most severely of the Nordic countries. Real GDP fell by 14.6 per cent from its peak in the first quarter of 1990 to its trough in the second quarter of 1993. This recession stands out among the industrialised OECD countries and was even worse than the Great Depression of the 1930s in Finland (Kiander and Vartia 1996: 72). In Finland, the biggest challenge to the welfare state in the early 1990s has no doubt been unemployment, which increased rapidly to an unprecedented level of nearly 20 per cent in 1993–1994. Only a few years earlier, in 1990, the unemployment rate was at 4 per cent and a labour shortage was quite seriously discussed. The increase in unemployment was reflected in housing expenditures in relation to households' incomes. In 1993, about 23 per cent of the disposable income of Finnish households was spent on housing on average. In 1990, the proportion of housing expenditures was about 18 per cent (Ministry of the Environment 1997: 39).

Another challenge, as in other industrialised countries, is the changing age structure of the population, which will lead to increased demands for social welfare services and, at the same time, a reduced potential for financing these services.[2] A new mix of welfare provision is needed to meet the demand for services of the ageing population. The aim is for people to manage at home for as long as possible, so that non-institutional services take priority over institutional ones.

Both challenges also affect housing policy. Concerning the age structure of the population, in Finland it has been found that living in poorly equipped dwellings is a problem of the elderly in particular. Living in dwellings lacking at least one of the basic amenities is also more common in rural areas compared with urban centres.[3] It is these 'granny's cottages' that have received assistance, for instance, in the form of repair grants. In urban areas, the lack of lifts in older apartment buildings is often an obstacle for the elderly who would otherwise prefer to live in their own home as long as possible. Reaching an agreement between the residents on new lifts is often a complicated matter for a Finnish housing company (preferences/cost sharing). The government has acknowledged the social value of lifts by including a subsidy for installing lifts in existing apartment buildings in the housing budget. Both rental and owner-occupied buildings qualify. These two examples reflect very specific aspects of housing policy. More general housing support systems are discussed below.

The recession led to a growing dependence on the state: the proportion of income transfers in the disposable income of households grew from one-fifth to one-third in five years. In fact, the Finnish welfare state managed to prevent large population groups from sliding into poverty because of large-scale unemployment, although social security benefits were cut at the same time. Contrary to expectations, the recession did not widen the income gap between households. Taxation and income transfers worked together to even out the drop in factor incomes. All income earner groups lost, more or less, equally (Heikkilä and Uusitalo 1997: 13).

Homelessness

Information on homelessness was collected for a Nordic research project. The number of homeless people per 1,000 inhabitants in the early 1990s was estimated at 0.4 in Iceland, 0.7 in Norway, 1.3 in Denmark, 1.5 in Sweden and 2.2 in Finland (Nordic Council of Ministers 1995: 35). The homeless are defined in the Nordic countries so as to include people living outdoors or in shelters, persons not able to leave an institution due to a lack of housing, and persons living temporarily with relatives or acquaintances because of a lack of housing. Notwithstanding the usual difficulties in measuring the extent of homelessness, the above estimates in the Nordic countries are all at low levels compared to European Union averages.

As the United Nations' International Year of the Homeless 1987 approached, the Finnish government introduced special measures to provide homes for the homeless. Funds were set aside in the state budget for grants and low-interest loans to allow the municipalities and approved private organisations to purchase dwellings for the homeless. A notable reduction in homelessness was achieved. Homelessness has now been reduced to about half what it was in 1987 (Table 14.1). Today, there are hardly any direct measures or special funds for housing the homeless. While there is less visible homelessness (people living in temporary shelters or doss houses), homelessness may have taken different forms and is therefore becoming more difficult to quantify (Kärkkäinen 1996: 76). These homeless people would be people who move around among relatives and friends because affordable small apartments are not available. The growing emphasis on non-institutional care and services will bring more pressure in future on the

261

Table 14.1 The homeless in Finland in 1987–1996

Year	A Outdoors or in shelters	B Cannot leave an institution because homeless	C Temporarily with relatives or friends because homeless	A + B + C Homeless individuals total	D Families living apart because of housing problems
1987	4,700	4,800	7,600	17,100	1,400
1990	3,600	3,700	8,000	15,300	800
1993	2,600	2,400	6,700	11,700	250
1996	1,700	2,100	5,800	9,600	350

Source: Housing Fund of Finland (1987–1996).

co-ordination between social welfare and housing provision to prevent de-institutionalisation from leading to an increase in homelessness.

Urbanisation

Iceland is the most urbanised of the Nordic countries with 92 per cent of the population living in urban areas in 1995. Finland, on the other end, has the lowest urbanisation rate at 65 per cent. Denmark (85 per cent, excluding Greenland and Faroe Islands), Sweden (83 per cent) and Norway (73 per cent) are in between. (Economic Commission for Europe 1997: 13–14.)

Urbanisation accelerated rather late in the Nordic countries. Migration from rural to urban areas increased the demand for housing mainly in the 1960s and 1970s. The intensity of urbanisation called for a more effective housing policy. Sweden had a 'million dwellings programme' in the late 1960s. Finland followed with a 'half a million dwellings programme' in the 1970s. The late beginning of the urbanisation process and its continuation today are a reason for the central role of subsidies for new housing production, which still prevails in Finnish housing policy. In the other Nordic countries, upgrading and modernisation of housing and urban renewal in general are relatively more important.

Income distribution

The success of the Nordic welfare model can be seen by comparing measures of inequality. Studies on income distribution distinguish the Nordic countries from the other OECD countries. For instance, when Gini coefficients were used to estimate the equality of income distribution, Finland, Sweden and Norway were found to have the lowest measured inequality (Finland 20.7, Sweden 22.0 and Norway 23.4 in 1986–1987). These countries were followed by Belgium, Luxembourg, Germany and the Netherlands (Gini coefficients in the range 23.5–26.8) (Atkinson *et al.* 1995: 46).

In Finland the stability of the income distribution during the recent recession has been rather surprising. The changes in the Gini coefficient have not been statistically significant during the recession years. In Table 14.2 it is shown how income transfers have effectively compensated for the more uneven distribution of factor incomes. The Gini coefficients for disposable incomes have therefore varied much less. Housing allowances are among the income transfers that contributed to the sustained stability in income distribution. The number of recipients of general housing allowances (mostly families with children) doubled from about 110,000 households in 1990 to over 220,000 in 1994 (Ministry of the Environment 1997: 47). Pensioners and students receive housing allowances from separate schemes. The total number of recipients (about 474,000) represented about 23 per cent of all households in 1994. Since then the number has decreased to about 20 per cent as the criteria have been tightened.

The future

The fiscal crisis has provoked public discussion about the future of the welfare state model also in Finland. The public discussion has borrowed neoliberal arguments from abroad. Disincentives created by social benefits, too high tax rates and excessive wardship of the state over the citizens have been the core claims of this critique. The proponents of the Nordic type of welfare state also claim that one must not bask in the glory of past achievements, but rather adjust the welfare state to new conditions and challenges (Uusitalo 1995: 6–7).

The programme of the present Finnish government states that an increase in employment and a decrease in the public debt to GDP ratio are essential to maintaining the welfare society. There has been a fairly common understanding of the necessity of restoring stability in state indebtedness to retain the basic thrust in the welfare provision systems. When the first phase of large cuts in public finances was over, a new and different tone, however, emerged in the discussion. The incentive effects of the social security systems and income transfers are now more prominent.

Table 14.2 Income distribution in Finland: Gini-coefficients in 1989–1995

	1989	1991	1993	1995
Factor incomes	39.3	39.7	45.1	46.4
Gross incomes	25.9	25.1	26.1	26.9
Disposable incomes	20.5	20.2	21.0	21.6

Source: Statistics Finland (1996, 1997).

Housing markets and housing policy

Social housing

State-subsidised social rental housing is made available in all the Nordic countries; its importance though varies as an instrument of housing policy. Social rental housing may be available to disadvantaged groups mainly or to the population in general. The Nordic countries can be divided along two main lines in their provision of social rental housing: there is a general model in Sweden and Denmark, and a more selective model in Iceland, Norway and Finland. The terms 'general' and 'selective' relate here to allocation of social rental dwellings. In the general model, dwellings are allocated to applicants usually according to a waiting list. In the selective model, housing allocations are based more directly on applicants' needs and income (Nordic Council of Ministers 1995). No income ceilings are applied in Sweden and Denmark. The waiting list system can be adjusted to take into account the applicants' housing needs. For instance, in Denmark, where social housing is owned by non-profit housing associations approved by the municipality, the municipality may itself assign every fourth vacant dwelling to solve urgent housing problems (Ministry of Foreign Affairs *et al.* 1996: 16). In Finland there are income ceilings but these reach well into the middle income households. Norway and Iceland have very small social rental housing sectors.

Income ceilings and other social criteria for obtaining housing in the social rental sector could accelerate visible segregation in the housing stock. The general model does not prevent segregation on its own either. The outcome depends on other policies applied to the design and construction of social housing – as well as tenant selection policies. In Finland there has been little difference, for instance, in the architectural style between social rental housing and owner-occupied housing in the multi-family sector. A Finnish housing company building can, in practice, have both social housing tenants and owner-occupiers because the ownership of an individual dwelling in a housing company is derived from the ownership of shares in the company giving entitlement to the possession of a particular dwelling.

Housing production

In Finland about 70 per cent of the housing stock is owner-occupied dwellings. The rest consists of private rental dwellings (17 per cent) and social rental dwellings (13 per cent). Finland has experienced an increase in the number of rented dwellings in the last few years in contrast to the trend in other European countries. In 1990, one-quarter of the dwellings were rented, in 1995 30 per cent.

After a strong construction boom, there was a drastic decrease in construction activity in the beginning of the 1990s. The number of completed new dwellings amounted to 13 dwellings per 1,000 persons in Finland in 1990. In 1995 the number declined to only 5 per 1,000. In other Nordic countries housing production has declined but the decline has been more stable (Table 14.3). Trends towards saturation have been seen in the Swedish and Danish housing markets. Finland is still at a later stage in the process. This could be inferred, for instance, by comparing the number of one-member households in the Nordic countries (Table 14.4).

Subsidies for social housing production

Finland

A state housing loan programme has been in operation in Finland since 1949. Loans have been granted from state funds for the production of owner-occupied housing and rental housing. The emphasis of the programme was first on owner-occupation and shifted only gradually to rental housing. State housing loans ('ARAVA' loans in Finnish) can be used to finance production of new housing as well as renovation and modernisation. Besides direct lending, the state has

Table 14.3 Housing construction in the Nordic countries: completed dwellings per 1,000 inhabitants

	1988	1990	1992	1994
Denmark	5.1	5.3	3.2	2.4
Finland	9.4	13.1	7.4	5.2
Iceland	7.4	6.9	6.1	6.3
Norway	7.2	6.4	4.1	4.1
Sweden	4.8	6.9	6.6	2.5

Source: Nordic Council of Ministers (1996).

Note
Denmark excludes Faroe Islands and Greenland.

Table 14.4 Population and households in the Nordic countries

	Year	Total population (000s)	Number of households (000s)	Share of one member households (%)
Denmark	1980	5,027	2,069	29
	1990	5,028	2,229	34
Finland	1980	4,708	1,782	27
	1990	4,927	2,037	32
Norway	1980	4,046	1,524	28
	1990	4,206	1,751	34
Sweden	1980	8,132	3,498	33
	1990	8,180	3,830	40

Source: Nordic Council of Ministers (1996).

supported housing production through various interest subsidies on private sector commercial loans.

In the late 1980s, it became necessary to dissociate the ARAVA loans from the state budget. Establishment of a separate fund for ARAVA lending was the solution. Another solution would have been to rely more heavily on interest subsidies to private sector financing. The Housing Fund of Finland was established at the beginning of 1990. The Housing Fund acts as a housing finance management body for the state, but it functions outside the state budget. The Fund, however, is not a separate juristic person; its obligations constitute obligations of the Republic of Finland.

Although it is separate from the state budget, the Housing Fund has not been immune to changes in public finances. During the recession, central government debt rose from 10 per cent of the GDP in 1990 to about 70 per cent in 1996. Since the Housing Fund's debt (i.e. housing bonds) is counted as part of the state debt, it was clearly necessary to consider other ways of raising funds for subsidised housing production. Instead of the sovereign debt market, the Housing Fund took advantage of the capital market integration in Europe and entered the mortgage-backed market (i.e. securitisation) in 1995. Securitisation means that the Fund utilises the credit of its high quality ARAVA loan portfolio as collateral for the funding instead of the 'name' of the Republic of Finland. The securities were issued on euromarkets through a special-purpose issuing vehicle established in Ireland. This transaction was the first securitisation sponsored by a European central government (Tulla 1996).

Because of securitisation, the state has been able to maintain an important role for state housing loans in Finnish housing policy. It has been possible to combine market-based funding with the implementation of the government's housing policy objectives, since the Housing Fund has its own assets in the form of the ARAVA loans. This funding arrangement has facilitated counter-cyclical measures and state support for housing production has been maintained at a considerable level during the recession. The share of state-subsidised housing production increased to 69 per cent of all housing starts in 1991–1995. During the housing market boom in 1986–1990 the state subsidised production accounted for 36 per cent of housing starts. In absolute numbers state subsidised production was 98,000 dwellings in both five-year periods (Ministry of the Environment 1996b: 66).

Sweden

The Swedish housing policy has been committed to the neutrality of support between tenures, and it is a little artificial to separate a social housing sector. The rental dwellings owned by municipality-controlled property companies account, however, for about 21 per cent of the housing stock.

Sweden and Finland have adopted very different approaches to financing the social rented sector. As described above, Finland has been able to maintain state housing loans despite the pressures on public finances because a new funding method was adopted. In Sweden, a state-owned mortgage company (the SBAB) was established in 1985 to grant second mortgages which borrowers had obtained before directly from the state. First mortgages came from private lenders. The distinction between first and second mortgage loans was abolished in 1992 when state guarantees on housing loans were introduced. The SBAB competes now with the private lenders. A state guarantee covers a proportion of a housing loan (the proportion of the former second mortgage). The state pays interest subsidies on the loans. The subsidies used to be fairly generous but have been subject to several cuts and modifications in recent years.

Denmark

Subsidised housing has constituted about 60 per cent of all new housing built in Denmark in the last five years (Ministry of Foreign Affairs *et al.* 1996: 26). Among the Nordic countries, Denmark has the lowest level of housing production (Table 14.3), and the demand for rental housing has been greater than the demand for owner-occupied housing which explains the large share of subsidised production. Social housing is about 18 per cent of the housing stock. As in Sweden, the state pays an interest subsidy on commercial loans for social housing.

Norway

Norway, like Finland, has a system of state housing loans. The Norwegian State Housing Bank (Husbanken) was established already in 1946. The Housing Bank has financed two-thirds of homes built in Norway since the mid-1940s. Social housing (i.e. municipality-owned rental dwellings) was, however, only 4 per cent of the housing stock in 1990 (18 per cent rented from a private landlord, 19 per cent co-operatively owned, 59 per cent in private ownership) (Ministry of Local Government and Labour 1996: 24). The co-operative sector provides housing to population groups which otherwise would have difficulties in obtaining housing. The Housing Bank used to finance only new housing production but as the production of new dwellings has become less important, finance has been made available for renovation and urban renewal as well. The lending volumes, interest rates and terms of payment are decided by the government as part of the state budget. Because of its direct involvement in lending, the state has been able to ensure a reasonable level of housing production in all circumstances. Financially this has been possible because of the strong economy and sound public finances.

Iceland

Social housing has been given more priority in recent years in Iceland. In the 1990s about 35 per cent of newly built housing has been social housing.

Homeownership dominates in Iceland more than in any other Nordic country, reaching 80 per cent. A clear difference from the other countries is that social owner-occupation represents the bulk of social housing. In recent years home-owners have faced difficulties in debt-servicing in Iceland. Social housing may become more diversified in future as demand grows for rental housing and co-operative housing (Sveinsson 1996: 215). Social housing is financed by the State Housing Board.

Iceland, Norway and Finland are countries where the state still has a significant role as a lender for housing purposes. In Finland, where public finances have been under pressure, a solution has been found to maintain the system of state loans. These financial pressures have a domestic origin – the severe economic recession as such – but they are also related to the external scrutiny of economic policies in the European Union in preparation for the Economic and Monetary Union. The two non-EU countries have not faced such drastic changes. Some kind of review of the functions between private and public lenders could be under way, especially in Iceland (Ministry for Foreign Affairs 1996: 53).

Financial market liberalisation

In the last ten years, fluctuation of house prices has been particularly strong in Finland. House prices increased about 60 per cent in real terms from 1985 to 1989. After the economic boom, house prices fell about 50 per cent in real terms from 1989 to 1993. Since then, prices have fluctuated in a moderate fashion. The volatility of house prices in relation to income has been commonly related to changes in financial market conditions. In 1986 interest rate controls were abolished, inducing a huge growth in credit. Growth in real disposable income of households fuelled demand while the real cost of borrowing was still very low. At the peak of the credit boom in 1989, acquisition of an average size dwelling cost a middle income family the disposable income of some three and a half years. Four years later, the acquisition required the disposable income of less than two years (Ministry of the Environment 1997: 39).

Credit expansion and house price increases followed the timing of deregulation: Denmark in the early 1980s, Norway and Sweden in the mid 1980s and Finland after the mid-1980s. Denmark, Norway and Sweden were able to phase deregulation of the domestic capital market together with the foreign exchange deregulation but in Finland all changes were implemented in a few years. This clarifies why house prices increased most in Finland. In Denmark, the timing of deregulation and tax reforms, and macroeconomic development as a whole, contributed towards stability in the housing markets. In Norway, Sweden and Finland the timing was less fortunate and instability increased in the housing markets (Nordic Council of Ministers 1994: 68–75). Financial market deregulation was followed by bank crises in these three countries. The public sector had to spend large sums to support the troubled institutions and the financial sector in general (Koskenkylä 1994).

The recession of the 1990s had a striking impact on households' finances and housing decisions in Finland. Many were forced out of owner-occupation because of indebtedness and potential owner-occupiers have postponed their home purchases. A regime shift took place in the Finnish housing market as a consequence of financial market liberalisation. Only after deregulation did house prices become responsive to real interest rates. The tax deductibility of interest payments on housing loans was restricted in Finland in 1992. The change took place during the economic recession and house prices fell further. In Sweden the tax changes took place a little earlier, but coincided there also with the economic recession.

Rent deregulation

Deregulation has been pursued in the private rental markets as well. New tenancy legislation came into force in Finland in May 1995. Rent-setting is now based on freedom of contract. The landlord and the tenant agree between themselves on the rent, lease period and the way rent is reviewed during tenancy. Before the reform, the government issued annual guidelines on reasonable rent levels and on acceptable adjustments to the rent levels. Under the new legislation, a reasonable level for rents is set subject to a market test without direct government intervention. The landlord or the tenant may request a court to consider the reasonableness of the rent in relation to market rents of similar rental dwellings.

Initial supply responses to the deregulation of rents have been quite strong in Finland. The Ministry of the Environment and Statistics Finland have followed market outcomes of deregulation regularly since 1991, when first steps of partial deregulation were taken. Between 1992 and 1996 an estimated net 50,000 private rental dwellings came on the market, reversing the trend towards more owner-occupation (Ministry of the Environment 1996a: 12). Transfers from owner-occupation to renting can happen easily in Finnish housing company buildings within the multi-family sector. Increases in market rents have been moderate so far, which has to do with the timing of the deregulation during the recession.

It is not yet clear whether the trend will reverse again towards owner-occupation, when the economy grows and disposable incomes increase. The underlying preferences favouring owner-occupation have not disappeared, not even among the young people whose unemployment rate exceeds the average. Saving for a home is, however, not as systematic or deliberate as before the recession (Niska 1996). This is reflected in reduced participation in the state-subsidised home savings scheme designed for first-time home buyers. One might predict that a good majority of people will still end up owner-occupiers at some stage. The difference is that the process of trading up in the housing market will probably begin when people are older.

In Sweden, private market rents have been more regulated than in Finland. A negotiation mechanism exists between tenants' and landlords' associations, and a rent tribunal determines reasonableness of rents on the basis of so-called use values of similar rental dwellings in the social rented sector. In new lease agreements the landlords and tenants can, however, negotiate directly on the rent level, and more flexibility is allowed in the agreements. In Norway, the basic principle is similar to that in Finland: the rent cannot be unreasonably higher than the average rent for similar rental housing. The municipalities can, however, regulate the rents in some parts of the stock if there is strong pressure on the rental market (Ministry of Local Government and Labour 1996: 50–51). In Denmark, rents may not be fixed freely and the rent levels do not reflect market conditions as much as in Finland. Recently, a commission has examined the rent legislation in detail and some deregulation is expected to take place.

In Finland, the deregulation of rents was motivated by a desire to expand the private rental sector so that it could become a true alternative to owner-occupation. Sweden and Denmark, with their relatively large social housing and co-operative housing sectors and a more universal approach to social housing, have offered more housing options than the Finnish housing market, and therefore they have not deregulated their rental market to the same extent as Finland. In Sweden, the question is more about the balance between the co-operative sector, private rental housing and social rental housing because, in contrast to Finland, owner-occupation is confined to single-family housing. A substantial transfer of stock from the public sector to the private sector could take place in Sweden if housing provision by local authorities becomes a matter for closer consideration (McCrone and Stephens 1995: 134). Norway has also a significant co-operative sector, 19 per cent in 1990 (Ministry of Local Government and Labour 1996: 39), which falls between owner-occupation and rental housing. The popularity of co-operative housing is increasing also in Finland. But in Finland this housing alternative was created only at the beginning of the 1990s and the total stock is still very small.

Relations between the housing markets and European integration have received hardly any attention in Finland so far. Labour mobility is, however, coming under focus more frequently in the discussion on adjustment mechanisms in the European Economic and Monetary Union, especially because mobility across European borders has been at a low level. Furthermore, the number of foreigners living in Nordic countries as a proportion of total population is low compared to the rest of Europe. Sweden, however, is an exception: foreign citizens accounted for 5.6 per cent of the population in 1990 (9.2 per cent of the population was born in a foreign country). Finland had the lowest proportion at 0.5 per cent (1.3). For Denmark and Norway, the proportions were 3.1 per cent (4.5) and 3.4 per cent (4.3) respectively (Fischer and Straubhaar 1996: 107). Some convergence in these proportions is likely to happen in the future. While migration across national borders may increase somewhat, occupational and regional mobility are relatively more important adjustment mechanisms to economic shocks in integrated markets. In this respect, better availability of private rental housing would be favourable for labour mobility.

Costs of housing subsidies

The costs of Swedish housing subsidies have been found to be particularly high in comparison with other European countries. In 1991 the cost was 4.1 per cent as a proportion of GDP (McCrone and Stephens 1995: 136). The high costs relate to the universality and non-discriminatory orientation of Swedish housing policies. Interest subsidies increased the costs rapidly, especially at the beginning of the 1990s when market interest rates rose. To curtail the increase in costs, first, tax relief for owner-occupiers was reduced. Second, the government cut interest subsidies and housing allowances. Recently, it has been proposed that the interest subsidy system should be replaced with lump-sum subsidies on new rental housing (*Bostadspolitik 2000* 1996). The emphasis is now more clearly on urban renewal in general rather than new production. On the other hand, Sweden still seems to be committed to tenure neutrality in its housing policy.

In Finland, the costs of housing subsidies amounted to about 1.8 per cent of GDP in 1991 (Table 14.5). Included in the costs are housing allowances, interest subsidies and tax relief for owner-occupiers (deductibility of interest on housing loans). The costs of housing subsidies increased during the recesssion in Finland because of the expenditures on housing allowances. Repair grants were increased to stimulate employment in the construction sector. At the same time, tax relief for owner-occupiers was reduced. When the effect from a fall in GDP is added, the outcome was that the costs of housing subsidies increased, reaching 2.3 per cent of GDP in 1993. Since then, the percentage has been falling to pre-recession levels. In Norway it has been estimated that in 1992 tax benefits amounted to 3.7 per cent of the GDP while direct housing subsidies were only 0.4 per cent of the GDP. Tax benefits have been reduced so that in 1994 they were 1.8 per cent of the GDP and direct benefits have increased to 0.6 per cent of the GDP (Ministry of Local Government and Labour 1996: 28).

Table 14.5 Housing subsidies in Finland (FIM millions)

	1990	1991	1992	1993	1994	1995	1996
Housing allowances	1,857	2,251	3,077	3,303	4,024	4,275	3,900
Interest subsidies							
– on private sector loans	604	715	842	821	572	783	855
– on state housing loans	1,694	1,761	2,511	2,547	2,767	2,482	2,300
Grants (e.g. on repairs)	119	75	97	349	410	359	561
Tax relief on housing loan interest	4,100	4,200	4,200	4,100	3,400	3,200	2,400
Total	8,374	9,002	10,727	11,120	11,173	11,099	10,016
– as a percentage of GDP	1.62	1.83	2.25	2.31	2.19	2.03	1.76

Source: Housing and Building Department, Ministry of the Environment.

Conclusion

A shift in emphasis from new production to renovation and urban renewal in general has also been visible in Nordic housing policies. Denmark and Sweden began to identify problematic housing estates in the 1980s (Ministry of Foreign Affairs *et al.* 1996: 22; Ministry of Industry and Commerce 1996: 54). In Finland, urban policy did not really become an issue until unemployment changed the social fabric of many housing estates in the early 1990s. At the same time, the Finnish experiences in the 1990s speak for a more pronounced counter-cyclical role of subsidised housing production.

European integration is expected to accelerate with the adoption of a single currency. In Finland, a series of discussion papers on the potential effects of the European Economic and Monetary Union on the Finnish economy has been commissioned from a group of experts. Although housing is not directly on the agenda, one can assume that there are potential effects on housing provision. Public finances are subjected to closer scrutiny today and in future – with or without EMU. Tax relief has been the first target of cuts in the Nordic countries, Norway included. Direct subsidies are currently under consideration in Sweden. Concerning the European Union, attention has been paid to the possible harmonisation of taxation at the Union level under a potential threat of tax competition (Julkunen 1997). Harmonisation would especially affect the Nordic member states of the EU which have financed their extensive public services and income transfers with high taxes. On the other hand, migration and labour mobility might have to be taken into account more actively in housing policies.

Acknowledgements

I have benefited from discussions with colleagues at the Housing and Building Department at the Finnish Ministry of the Environment. The views are, however, those of the author only. Terry Forster kindly helped me with the English language.

Notes

1 Finland has a population density of 15 inhabitants per square kilometre, Sweden 20, Norway 13, Iceland 3 and Denmark 121 (Economic Commission for Europe 1997: 13–14).
2 Iceland, however, has Western Europe's lowest proportion of people over 65 years of age (Sveinsson 1996: 217).
3 A well-equipped dwelling is defined in Finland as one having the following basic amenities: piped water, sewer, hot water, flush toilet, bathing facilities and either central heating or direct electric heating. Nearly 90 per cent of the population have these amenities (Ministry of the Environment 1997: 33).

References

Atkinson, A.B., Rainwater, L. and Smeeding, T.M. (1995) *Income Distribution in OECD Countries*, Social Policy Studies No.18. Paris: OECD.

Bostadspolitik 2000 (Housing Policy 2000) (1996) Slutbetänkande av bostadspolitiska utredningen, SOU 156. Stockholm: Inrikesdepartementet.

Economic Commission for Europe (1997) *Annual Bulletin of Housing and Building Statistics for Europe and North America*. Geneva: United Nations.

Fischer, P.A. and Straubhaar, T. (1996) *Migration and Economic Integration in the Nordic Common Labour Market*, Nord 2. Copenhagen: Nordic Council of Ministers.

Heikkilä, M. and Uusitalo, H. (eds) (1997) *Leikkausten hinta* (The Price of Cuts). STAKES Raportteja 208, Gummerus: Saarijärvi.

Housing Fund of Finland (1987, 1990, 1993, 1996) *Housing Market Survey*. Helsinki: Housing Fund of Finland.

Julkunen, R. (1997) *Suomalainen sosiaalipolitiikka Euroopan talous- ja rahaliitossa ja matkalla sinne* (Finnish Social Policy on the Way to the Economic and Monetary Union), valtioneuvoston kanslian julkaisusarja 19. Helsinki: Prime Minister's Office.

Kärkkäinen, S.L. (ed.) (1996) *Homelessness in Finland*, STAKES (National Research and Development Centre for Welfare and Health). Gummerus: Jyväskylä.

Kiander, J. and Vartia, P. (1996) 'The Great Depression of the 1990s in Finland', *Finnish Economic Papers* 9, 1: 72–88.

Koskenkylä, H. (1994) 'The Nordic Banking Crisis', *Bank of Finland Bulletin* 68, 8: 15–22.

McCrone, G. and Stephens, M. (1995) *Housing Policy in Britain and Europe*. London: UCL Press.

Ministry of the Environment (Finland) (1996a), *Uusien vuokrasuhteiden vuokrat* (Rents in New Tenancies). Helsinki: Edita.

Ministry of the Environment (Finland) (1996b) *Sustainability as a Challenge*, Finland's National Report to the Second United Nations Conference on Human Settlements (Habitat II). Helsinki: Ministry of the Environment.

Ministry of the Environment (Finland) (1997) *Housing Indicators*. Helsinki: Edita.

Ministry for Foreign Affairs (Iceland) (1996) *Settlement and Society in Iceland*, National Report to the UN Conference on Human Settlements (Habitat II). Reykjavik: Ministry of Foreign Affairs.

Ministry of Foreign Affairs, Ministry of Housing and Building and Ministry of Environment and Energy (Denmark) (1996) *The Danish National Report to Habitat II*. Copenhagen: Ministry of Foreign Affairs.

Ministry of Industry and Commerce (Sweden) (1996) *Shaping Sustainable Homes in an Urbanizing World*, Swedish National Report for Habitat II, Government Official Reports 48. Stockholm: Fritzes.

Ministry of Local Government and Labour (Norway) (1996) *From Reconstruction to Environmental Challenges*, Norway's National Report to the UN Conference on Human Settlements (Habitat II). Oslo: Ministry of Local Government and Labour.

Niska, A. (1996) 'Young People, Recession and Housing Changes in the 1990s', paper presented at the ENHR Housing Research Conference, Helsingor, Denmark, 26–31 August.

Nordic Council of Ministers (1994) *Kreditrisker och betalningsproblem i bostadssektorn* (Credit Risks and Payment Problems in the Housing Sector). TemaNord 666, Copenhagen: Nordic Council of Ministers.

Nordic Council of Ministers (1995) *Offentlige og allmennyttige utleieboliger i Norden* (Public and Non-Profit Rental Housing in the Nordic Countries). TemaNord 625, Copenhagen: Nordic Council of Ministers.

Nordic Council of Ministers (1996) *Yearbook of Nordic Statistics*, vol. 34. Copenhagen: Nordic Council of Ministers.

Solem, E. (1977) *The Nordic Council and Scandinavian Integration*. New York: Praeger.

Statistics Finland (1996) *Income Distribution Statistics: Income and Consumption 21*. Helsinki: Statistics Finland.

Statistics Finland (1997) *Income Distribution Statistics: Income and Consumption 12*. Helsinki: Statistics Finland.

Sveinsson, J.R. (1996) 'Main Trends of Icelandic Housing in the 1980s and 1990s', *Scandinavian Housing & Planning Research* 13: 215–220.

Tulla, S. (1996) 'Securitisation and Finance for Social Housing: New Developments in Finland', *European Mortgage Review* 2 (May), Council of Mortgage Lenders.

Uusitalo, H. (1995) *The Future of the Finnish Welfare State*, STAKES (National Research and Development Centre for Welfare and Health). Gummerus: Jyväskylä.

15

EUROPEAN INTEGRATION AND THE EAST-CENTRAL EUROPEAN 'OUTSIDERS'[1]

Iván Tosics

There are at least two obstacles to analysing the problems of an extension eastwards of the EU from a housing policy point of view. One is the general lack of information about the housing component of European integration. The second is the lack of reliable comparable information on housing, not only showing the differences between Western and Eastern European countries, but between the eastern countries themselves. The situation has somewhat improved in the last years in relation to the second problem with the completion of the East-Central European Regional Housing Indicators project (funded by USAID with additional help from UNCHS/Habitat and the ECE) from which comparable information is available on the relative housing situation of the individual Central and East European countries.

In this chapter we address three important questions from a housing policy point of view:

- What are the pros and cons for the EU of an extension to the east?
- What are the expectations of and possible effects on the eastern countries regarding joining the EU?
- What kind of differences exist between the countries of central and east Europe regarding their desire and ability to join the EU?

We have to emphasise that at this moment we cannot give well-documented answers on these questions. However, we can use the available (limited) empirical data, and, just as importantly, place these questions into a coherent framework.

The Central and Eastern European region: background

Classifying the countries of the Central and Eastern European region

As the consequence of the dramatic changes (the formation of nation-states) of recent years the number of countries in the Central and Eastern European region has increased greatly. According to some calculations there are more than forty 'transitional' countries (on their way from planned to market economy) all over the world, the majority of which are in Europe or in the adjacent areas. We cannot deal with this number of countries but there is no need to do so as the problem of EU integration is limited to a much smaller circle of countries. As a working hypothesis (taken from Pickel and Pickel 1995) we can establish four groups of countries which can be regarded as being in different positions from the point of view of EU integration.

The group of post-Soviet states (referred to later in this chapter as 'Eastern East European' or EEE countries) has the least chance of joining the EU in the short term. However, this does not mean, as we will see, that they do not have such hopes in the longer term.

The North Eastern European (NEE) countries, although having already strong ties to Western Europe, especially the Nordic countries, are in a special position as a consequence of the special interest Russia still expresses towards any idea of this group joining the EU or NATO.

The group of South Eastern European (SEE) countries is a very mixed group, consisting of countries in very different positions regarding their economic and political development. Some of these countries have definite political aims to join the EU already in the first round, while others do not aim for this in the short run or do not see it as realistic.

The Central Eastern European (CEE) countries consist of the V-4 (Visegrad) countries plus Slovenia as the least war-affected Yugoslavian successor state. This

Table 15.1 The classification of countries in the Eastern European region from the point of view of EU integration

Central Eastern European states	South Eastern European states	North Eastern European states (Baltic countries)	Post-Soviet states
Poland	Romania	Latvia	Euro-Russia
Czech Republic	Bulgaria	Lithuania	Belorussia
Slovakia	Macedonia	Estonia	Ukraine
Hungary	Albania		Armenia
Slovenia			Georgia

Source: Pickel and Pickel (1995).

Note
Croatia and Yugoslavia could be classified into the South Eastern European group.

group of countries forms the CEFTA (Central European Free Trade Area) group which aims to decrease inter-country barriers to trade. The CEE countries are now the 'neighbours' of the EU and have already built up strong relationships with some of the institutions of the EU.

The idea of a 'buffer zone' on the eastern border of the EU (see Wallace *et al.* 1995) is illustrated in Figure 15.1. The four countries belonging to this zone consist of those which are in a geographical sense in between the EU and the successor countries of the Soviet Union. From the beginning of transition it was clear that countries with seventy years of socialist-communist heritage need more time to adjust to the democratic and market-based model than countries with 'only' forty years of that heritage. Furthermore, within the latter group, the EU neighbour countries are in a special position.

It was an early recognition of the EU that the potential problem of a mass exodus from the east to the west could be handled (only) by establishing a new 'iron curtain' on the eastern border of the EU. A much better solution is to create relative stability in the neighbouring countries and try to convince them that, in the hope of early NATO and EU membership, they should contribute to the guarding of the eastern border of the EU (see Wallace *et al.* 1995: 41) The special handling of the buffer zone means that the CEE sub-region is in a natur-ally advantageous position with regard to joining the EU.

The East European model of housing policy

In earlier articles (e.g. Hegedüs and Tosics 1996) we have described in more detail the logic and the historical development of the housing policies of the Central and East European countries and we called this special form of housing policy the East European Housing Model (EEHM). The main features of this type of housing policy were direct state control over housing built by the state, co-operatives and saving banks (allocated partly on the merit principle, partly according to social principles), and indirect state control over private forms of housing connected with the control over households' income. There were some changes in the development of this model in the different countries (described as 'cracks' in the originally unified model). The 1970s and the first part of the 1980s were the 'best years' of this kind of housing policy in most Central and East European countries, showing a relatively high level of new construction (almost 10 new units per 1,000 population per year in Hungary).

In our analysis we showed that this system functioned with huge problems (inefficiencies and inequalities) regarding the allocation of subsidies, the use of investment in the building industry, the urban consequences of new construc-tion, etc. Even in the period of relatively high budget expenditures on housing, EEHM was a quite ineffective way of allocating this money. Over-centralised, over-controlled housing policies were developed, which, being almost totally dependent on the state budget, had to be changed at the first signs of budget difficulties.

Figure 15.1 North Eastern, Central Eastern, and South Eastern European states
Source: Pickel and Pickel (1995), Wallace *et al.* (1995).

In some of the countries the changes came some years before the change in the political system. Slovenia, Hungary and Poland started to reform their housing policies in the 1980s, decreasing control over the private sector and changing the subsidy system to a more balanced one, across the different housing forms. In most of the countries of the region, however, the really substantial changes came in the early 1990s, mainly in the form of massive give-away privatisation of the public rental housing stock. As a result, the control of the state decreased to a minimum level, both regarding political-legal regulations and the share of the non-private housing forms. Housing policies of the countries of the region became even less regulated and less state-controlled than those of the Western European countries.

Evaluation of the political and economic situation in the East European countries

In the last few years extensive attitude surveys have been carried out to get an overview of the opinion of the population regarding the most important processes of change. Figure 15.2 shows the outcome on two questions, the first of which measured the degree of satisfaction with the way democracy is run in the given country, while the second measured the subjective evaluation of economic development by the people. (The source is the Central and Eastern Eurobarometer 3 and 4, from 1992 and 1993, having approximately 1,000 random samples from each of the countries, see Pickel and Pickel 1995: 1–2.)

The results show a very differentiated picture within the CEE and SEE sub-regions, while the Baltic countries and the EEE group seems to be much more homogeneous. The authors conclude:

> We draw the conclusion that the long-term perspective of political stabil-
> ization is built up on the short-term perspective of the current econom-
> ical improvement, or properly speaking, on the appearance of a currently
> better economical situation.
>
> (Pickel and Pickel 1995: 7)

While not denying the existence of such a correlation, there are also other determinants, in our opinion, influencing the level of satisfaction with the new political system. One of those is the past experience of each country with any kind of political freedom. The huge difference, for example, between the Czech Republic and Poland on the one hand and Hungary on the other regarding the satisfaction with the way democracy is run, cannot be fully explained on the basis of satisfaction with economic development. The very positive reaction of Czechs and Poles is obviously connected to the fact that political freedom in these countries before 1989 was much more limited than in Hungary.

There is another important factor to be measured, namely the differences

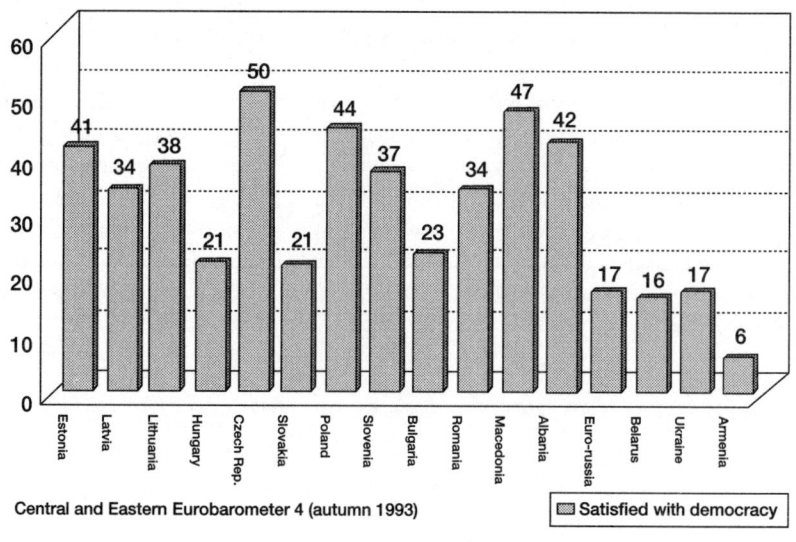

Central and Eastern Eurobarometer 4 (autumn 1993)   ▦ Satisfied with democracy

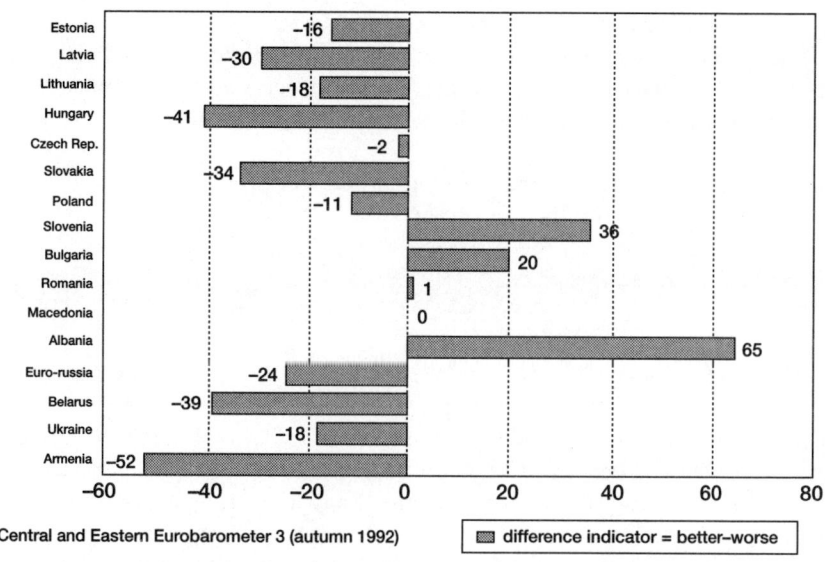

Central and Eastern Eurobarometer 3 (autumn 1992)   ▦ difference indicator = better−worse

Figure 15.2 Satisfaction with the way democracy is run and anticipation of economic development in Eastern Europe

Source: Chart 15.3 and Chart 15.4 from Pickel and Pickel (1995).

among the Central and East European countries regarding the desirability of joining the EU. Table 15.2 summarises attitudes to integration.

The first part of the table is based on the Central and Eastern Eurobarometer 3 (autumn 1992), relating to attitudes to joining the EU among CEE countries. The second half of the table is based on Eurobarometer 38 and shows the percentage of EU population accepting the citizens of the different Eastern and Central European countries.

The comparison of the two data-sets shows that the desire in the population to join the EU is the highest in the CEE and SEE countries, while data on the acceptance of the countries by the population of the EU show preferences for the CEE and Baltic countries. The group of EEE countries is at the bottom of both preference-lists. We should not accept the results of this type of evaluation uncritically. Apart from the methodological problems (scale differences, averages of countries of very different size) there is also a theoretical problem: the rejection of integration (or wanting to postpone it) can be based on very different attitudes: the Swiss or Danish type of arguments are completely different from the arguments of those who believe in the necessity of keeping a strong Eastern Europe and not integrating into the Western world. Another aspect is the

Table 15.2 Attitudes towards integration

	Attitudes to the time-point of integration into the European Union							EU population: acceptance of the given country (per cent)	
	Never (1) (%)	Over 10 years (2) (%)	10 years (3) (%)	In 5 years (4) (%)	Now (5) (%)	Average of 1–5	Group average	% accepts	Group average
Hungary	2	3	4	31	40	4.30		69	
Czech Rep.	3	6	10	45	28	3.97		57	
Slovakia	2	8	9	43	33	4.02		46	
Poland	4	5	7	23	44	4.18		68	
Slovenia	1	3	8	44	39	4.23	4.14	42	56.40
Bulgaria	1	4	6	29	30	4.19		52	
Romania	1	6	2	13	66	4.56		52	
Macedonia	4	3	2	15	70	4.53		36	
Albania	1	19	25	32	12	3.39	4.17	35	43.75
Estonia	3	10	10	29	25	3.82		55	
Latvia	2	7	12	32	25	3.91		53	
Lithuania	1	17	21	36	13	3.49	3.74	53	53.67
Russia	4	21	21	22	2	2.96		47	
Belarus	5	24	20	26	2	2.95		42	
Ukraine	3	19	23	27	2	3.08		31	
Armenia	5	23	18	25	7	3.08	3.02	44	41.00

Source: Pickel and Pickel (1995).

mixture of personal or patriotic wishes with political realities. However, there are no more detailed data available which would make it possible to refine the results.

The housing situation in the East-Central European countries

In this chapter we give a short overview on the housing situation of the East-Central European countries (and where possible, also in comparison with data on the EU countries). We do not have all the data necessary to show the differences among the four groups of countries (in particular data for the EEE group are missing) thus the following tables are only illustrations. Our results (taken from the USAID funded database, MRI, 1996) – concentrating on the first three sub-regions – show the need to continue the work on a system of reliable regionally relevant indicators in order to understand the specifics of the transition process in the East-Central European countries. The analysis and evaluation of data in this section is based mainly on Hegedüs *et al.* (1996).

The new tenure structure and basic quantitative/qualitative indices

From Table 15.3 it can be seen that there is no common tenure structure pattern in the region and there are huge differences even within countries belonging to the same group (cf. the difference within the CEE sub-region, i.e. between the Czech Republic, Poland and Slovakia on the one hand and Hungary and Slovenia on the other). As a consequence of massive privatisation in some countries the share of the owner-occupied sector became very high, much higher than 'necessary' taking the high proportion of families below the poverty line into account. By comparison, EU data on homeownership show a spread between 38 per cent and 81 per cent which is a smaller range than that between the countries listed in the table. The share of the rental sector is higher in the EU countries than in the transitional countries, mainly due to the very big difference in the share of the private rental sector.

In most of the EU countries housing shortages on the national level disappeared as early as the 1970s and the number of housing units exceeded the number of households. This was not the case in the East-Central European countries. Migration from rural to urban areas was still strong and in most of the countries huge state-subsidised new housing construction programmes had to be initiated in the 1970s. As a result (see Table 15.4), urban housing shortage is not a general problem of the region any longer in the mid-1990s (except for some of the countries, notably Poland, and most of the successor states to the Soviet Union). However, the densities shown in Table 15.4 highlight the special situation of the countries of the SEE sub-region: the number of flats equals or even exceeds the number of households, but the size of the flats (number of rooms and floor area) is very small, resulting in high density indices.

Table 15.3 The tenure structure in Eastern Europe around 1994 (%).

	Public rental	Private rental	Other rental	Owner occupied	Other (e.g. co-op)
Czech Rep.	27.6	4.7	6.1	42.2	19.4
Hungary	13.0	1.0	NAV	86.0	0.0
Poland	25.4	5.2	13.5	41.7	14.2
Slovakia	26.0	0.5	0.5	51.6	21.4
Slovenia	8.9	3.4	0.0	87.7	0.0
Total CEE	23.1	3.9	8.2	52.0	12.8
Albania	2.0	NAV	0.0	98.0	0.0
Bulgaria	6.8	3.2	0.3	89.7	0.0
Croatia	10.6	3.7	NAV	84.5	1.2
Romania	7.8	3.0	0.2	88.9	0.2
Total SEE	7.6	3.1	0.2	89.1	0.3
Estonia	56.0	5.0	3.0	30.0	6.0
Latvia	54.0	5.0	2.0	39.0	0.0
Lithuania	12.9	8.5	0.0	78.6	0.0
Total Baltic	36.4	6.5	1.3	54.4	1.3
Total transition countries	18.6	3.8	5.9	64.2	7.5
Total EU	14.1	21.2	2.4	58.7	3.6

Sources: For data on the transitional countries, MRI (1996); for the EU data, EU (1993); there are fourteen countries included (the twelve former EU members with Austria and Sweden).

Note
NAV means that data are not available. The individual country data are weighted by the number of dwellings to get the regional averages.

The development of infrastructure showed a special pattern in the socialist period: the limited amount of capital devoted to infrastructure was concentrated on the cheaper sectors (such as water supply) as opposed to the more expensive ones (such as sewage systems). As a result a big 'infrastructure scissor' developed in all of the transitional economies: the piped water system is much more developed than the piped sewage system. 'Second-best' methods, such as septic tanks (which needed private instead of public investment and were cheaper in terms of the immediate investment costs, even if they had serious environmental consequences in the longer run), were widely used to compensate for the lack of sewage systems, making possible the relatively high share of flats with bathrooms.

EU data show a substantially better situation in the number of dwellings and their size/quality, compared to the situation in the transitional countries.

Table 15.4 Basic quantitative and qualitative aspects of the housing stock 1994

	Housing units per 1,000 people	Household per dwelling	Square metres per person	Person per room	Piped water (% of flats)	Bath or shower (% of flats)
Czech Rep.	397	1.01	25.5	1.04	96.9	90.9
Hungary	385	0.99	32.1	0.92	82.9	85.9
Poland	296	1.06	18.2	1.02	84.2	71.5
Slovakia	334	1.00	21.9	1.14	91.8	88.9
Slovenia	338	0.95	19.0	1.33	97.4	86.6
Total CEE	330	1.03	22.4	1.02	87.4	79.6
Albania	219	1.00	8.0	2.70	57.5	54.0
Bulgaria	405	0.88	16.7	1.00	*83.4*	*44.7*
Croatia	336	0.98		1.10	*86.2*	75.7
Romania	341	0.95	17.4	1.19	*53.6*	48.2
Total SEE	344	0.94	16.6	1.21	65.3	50.9
Estonia	410	1.03	32.0	1.18	*89.7*	*73.2*
Latvia	370	1.13	20.9	1.21	*76.6*	*65.9*
Lithuania	329	1.06	19.7	1.30	*58.9*	*53.8*
Total Baltic states	358	1.08	22.8	1.24	71.7	62.2
Total transition countries	337	1.00	20.4	1.11	78.4	68.2
Total EU countries	428		34.7	0.6		93.3

Notes
a   Numbers in italic are from UN (1995).
b   Data for Lithuania exclude private ownership dwellings.
c   The individual country data are weighted by the number of dwellings to get the regional averages.

Affordability issues: 'rent to income' and 'housing cost to income' ratios

Social issues are extremely important in the process of transition. The measurement of affordability was an important part of the original set of key indicators (established by UNCHS/Habitat and the World Bank), but the indicator 'rent to income' did not take the total housing-related cost into consideration. In the socialist countries both rents and services/utilities related to housing were heavily subsidised. In the transition period it was easier for the state and the local governments to cut the subsidies on services/utilities than on rents. For example, the prices of energy, water and sewage, and garbage collection were increased in many of the countries more or less to world market level.[2] Thus the indicator

Table 15.5 Housing expenditures (rents and utilities) to income in the public rental sector in 1990 and 1994

	Rent to income (%)		*Utility expenses to income (%)*		*Total housing expenses to income (%)*	
	1990	*1994*	*1990*	*1994*	*1990*	*1994*
Czech Republic	2.7	3.1	3.9	7.8	6.6	10.9
Hungary	5.0	3.8	5.0	19.7	10.0	23.5
Poland	1.0	1.8	5.0	11.1	6.0	12.9
Slovakia	5.0	5.3	7.2	14.7	12.2	20.0
Slovenia		5.2		9.1		14.3
Total CEE	2.4	2.7	5.0	12.4	7.4	15.1
Albania	1.0					
Bulgaria	15.2	1.3	6.8	6.8	22.0	8.1
Croatia						
Romania	8.3	0.2	8.3	9.5	16.6	9.7
Total SEE	10.4	0.5	8.0	9.2	18.4	9.7
Estonia		4.2		9.9		14.1
Latvia	0.8	1.8	1.5	9.1	2.3	10.9
Lithuania		1.5		17.5		19.0
Total Baltic states	0.8	1.7	1.5	13.9	2.3	15.6
Total transition countries	5.1	1.9	5.8	11.4	10.9	13.3

'total housing expenditures to income' (column 3 in Table 15.5) shows better the changes than the indicator 'rent to income' (column 1 in Table 15.5).

The average total housing expenditures to income are still lower than in most of the EU countries. However, as a consequence of the much lower level of household incomes, even a 20–25 per cent housing expenditure ratio causes hardship to many families. This also means that local and/or central governments have problems in increasing the low level of rents. The fact that utility price increases are 'crowding out' the possibilities of rent increases has serious consequences for the chances of renovating/modernising the run-down multi-family housing stock.

It must be mentioned that the indicator 'median house price to income' ratio (the ratio of the median free-market price of a dwelling and the median household income) is calculated from the total housing stock, not just new units. The price of newly constructed units is still very high compared to income all over the region.

Because of the major restructuring now under way in East-Central Europe some of the relations known from economics take a special form. It is a well-known fact that house prices (even if subsidised and in some countries

administratively controlled) were in the socialist system very high compared to incomes in the East-Central European region. Since the collapse of the socialist housing model, housing prices, both for newly constructed units and for real-estate
transactions, are free to be set by market forces. As an additional fact, data show
that countries are experiencing a major decrease of new housing output (see Table
15.7). These facts, however, have an unexpected relation to the change in housing prices. Despite the fact that in the centrally planned housing systems housing
price inflation was much higher than the consumer price index (CPI), in the last
few years, when price setting became free and construction dropped dramatically,
house price inflation – as shown in Table 15.6 – dropped as well, below the level
of the CPI. The median 'house price to income' ratio has dropped from a level
around 6·7 in 1990 to 4.4 in 1994 (within capital cities the drop was from 9.3 to
6.8).

Possible explanations for this paradoxical situation, which relate closely to transitional issues, could be as follows:

- housing ceased to be the major and safest form of savings, as there are other
  legal possibilities
- privatisation and restitution freed an additional supply of housing, which
  pushed down prices

Table 15.6 Housing price-to-income ratio and house price inflation in the owner-
occupied sector

	House price to income (ratios)		Real house price inflation (%)
	1990	1994	1994
Czech Republic	NAV	5.6	2.0
Hungary	5.9	5.7	−4.0
Poland	NAV	3.2	−9.0
Slovakia	4.0	5.6	−6.8
Slovenia	NAV	7.0	−2.7
Total CEE		4.3	−6.2
Albania			1.2
Bulgaria	25.0	5.8	−1.9
Croatia			
Romania	16.6	4.7	−30.0
Total SEE		5.0	−17.7
Estonia	NAV	3.6	−25.0
Latvia			
Lithuania	NAV	3.2	−4.7
Baltic states		3.3	−10.5
Total transition countries		4.4	−10.4

Table 15.7 Measures of new investment into the housing sector

	Housing investment to GDP			New construction per 1,000 population			Population change	Household formation/1,000 pop.	Size of new units in square metres	
	1990	1994	1994/1990	1990	1994	1994/1990	1994 to 1990 (%)	1990–1994 annual change	1990	1994
Czech Republic	3.2	2.4		4.3	1.8	42%	0.33	0.00244	77	87
Hungary	4.2	2.8	67%	4.2	2.1	49%	-0.94	0.00080	90	101
Poland	5.2	1.8	35%	3.5	1.9	54%	1.04	0.00066	77	89
Slovakia		0.2		4.9	1.3	27%	1.55	0.00042	82	94
Slovenia	4.2	3.0	71%	3.9	2.8	72%	1.22	-0.00038	93	102
Total CEE	4.2	2.0	48%	3.9	1.9	49%	0.67	0.00091	80	92
Albania	1.4			3.9	3.7	95%	0.09	NAV	36	
Bulgaria	1.5	0.8	53%	3.0	1.1	33%	-2.79	NAV	72	84
Croatia	3.2	2.0	63%	3.9	2.0	51%	0.00	NAV	80	84
Romania	0.9	0.7	78%	2.1	1.6	76%	-0.35	NAV	47	66
Total SEE	1.4	0.9	64%	2.6	1.6	62%	-0.81	NAV[a]	61	72
Estonia		0.2		4.8	1.3	27%	-4.13	-0.00481	62	82
Latvia	4.5	1.6	36%	5.0	0.3	6%	-3.79	-0.0074	61	79
Lithuania		2.7		6.0	1.8	30%	0.43	-0.00034	64	101
Total Baltic states	4.5	1.8	40%	5.4	1.2	22%	-1.89	-0.00353	63	92
Total transition countries	3.2	1.6	50%	3.6	1.8	50%	-0.02	0.00029	73	86
Total EU countries					5.0					92

Note
a There is no reliable data available for the number of households in 1994 in any of the SEE countries.

- because of decreasing GDP (and real income) households save more and do not invest in housing
- housing markets function better, residential mobility is increasing, restrictions on multiple ownership are eliminated
- the private construction sector is producing housing with characteristics sought after by those who previously were most constrained in their choices.

Yet the 'house price to income' ratio in the capital cities is still high when compared to that found in other comparable cities.

Changes in housing output

Production and investment both fell sharply over the period 1990–1994. Investment in housing relative to GDP fell from 3.2 per cent to 1.6 per cent (an average fall of 50 per cent) and ranged from a 65 per cent decline in Poland to only 22 per cent in Romania.

Output in physical terms also fell from 3.6 dwellings per 1,000 people in 1990 to 1.8 per 1,000 in 1994, a drop of 50 per cent, with the steepest drops in the Baltics. (EU countries have an almost three times higher new construction rate, despite the already better housing situation.) Notwithstanding these drops in physical production, in a number of countries in the region (notably in the Baltic countries and in Slovenia) the number of households fell over the same period, while in the CEE countries the increase in the number of households was smaller than the increase in the housing stock. As a consequence, the net number of dwellings per household actually increased in the region, i.e. the household/ dwelling indicator decreased from 1.02 to 1.00.

The opening up of housing markets to private construction activity has resulted in a significant increase in the size and quality of dwellings being produced, generally in response to effective demand in parts of the market that were restricted as a measure of policy in the former system. From 1980 to 1990, average unit sizes of newly built dwellings had increased from 62 to 73 square metres, an increase of about 18 per cent. But in the subsequent four years the average size of new units being produced grew to 86 square metres – an average increase of 22 per cent (and a range of growth between 5 and 58 per cent within the region). One measure of the extent of this 'upmarket' move in production comes from comparing production in Eastern Europe to that of Western Europe, where the average size of newly built dwellings in 1994 was 92 square metres – a difference of less than 10 per cent despite the fact that officially measurable incomes in Eastern Europe were only about one-eighth as high as those in Western Europe.

Housing finance

The role of housing finance within the overall financial system, which was generally small to begin with, has generally lessened over the early years of the transition. This is partly due to the high nominal and real interest rates, partly to the lack of competition and other institutional problems. The Housing Credit Portfolio, the share of housing loans among the assets of the banking system, was below 9 per cent – a very low level even compared to countries at considerably lower levels of economic and financial development. As a consequence, the 'credit to value' ratio is also very low, only 12 per cent.

In the socialist housing system there was only one single bank to make all housing loans. By 1994, conditions were little changed: the median share of housing loans made by the largest lender in the sector had fallen only to 83 per cent. At the same time, however, lending practices were being rationalised throughout the region. The interest rate spreads between lending rates and deposit rates in housing banks became positive in most countries – even so, these were lower than spreads available on other types of lending, including that for central government financial obligations. This situation considerably dampened the willingness of even the housing banks to extend loans for housing. Nevertheless, according to Hegedüs *et al.*:

> Other innovations, however, offered more promise in changing the incentives for banks to lend and households to borrow. These included the introduction of indexed mortgages such as the Deferred Payment Mortgage in Hungary, the introduction of a mortgage banking system which relies in part on issue of mortgage bonds to raise funds in the Czech Republic, and the introduction of a variant on the German Bausparkasse system in the Czech Republic and Slovakia.
>
> (Hegedüs *et al.* 1996: 126)

The housing situation of East-Central European countries in a comparative view

At the end of our overview of the housing situation of the countries of East-Central Europe we compare some of the data of the twelve Eastern European countries to those of countries and cities in Western Europe,[3] and countries and cities with similar levels of economic development throughout the world.[4]

Comparisons of key indicators at the beginning of the transition period suggest that the housing situation of most of the East-Central European countries (especially of those of the CEE sub-region) is surprisingly good compared to other countries of similar economic development. At the same time, the housing situation is substantially worse than in the EU countries. Consequently, the issue of housing shortage which is frequently alleged within the region is likely to be substantiated only on the basis of either inappropriate comparisons to Western Europe, with incomes eight times as high, or expectations which have been conditioned by decades of low and distorted prices.

On the other hand:

Table 15.8 Housing finance indicators

	Housing credit portfolio		Credit to value	Housing loan concentration	Housing loan to one year deposit	
	1990	*1994*	*1994*	*1994*	*1990*	*1994*
Czech Republic	5.2	3			−3.0	6.0
Hungary	16.8	8.4	10.4	80	−0.1	8.1
Poland	4.1	19	17.9	96	−115.6	9.5
Slovakia	5.8	4.2	10	98	−1.5	5.2
Slovenia	25.5	14.9	37.2	42	NAV	2.8
Total CEE	7.0	13.6	16.5	91.4	−69.7	8.2
Albania		8.9		94	NAV	8.3
Bulgaria		1.9	5.5	85	−6.0	−8.0
Croatia				74		
Romania		0.6	3.7	85	−4.0	18.0
Total SEE	2.2	1.7	4.2	84.4	−4.5	10.7
Estonia		0			0.0	0.0
Latvia		2.8			NAV	50–100
Lithuania		2.7	16	40	−1.0	9.0
Total Baltic state		2.2	16.0	40	−0.7	20.8
Total transition countries		8.7	12.2	85.4	−42.5	9.9

Note
Weighted by the number of population in 1994.

there is ample evidence that there are significant problems in the distribution of housing. Relative to market-oriented economies, there is far less correspondence between household income levels and housing quality/quantity outcomes; and many more large households occupy small dwellings and small families large dwellings than is the case in market economies. The level of upkeep and maintenance of the existing multi-family housing stock is far below that of market economies, there is a significant 'deferred maintenance', relative to the economically necessary normal maintenance cycles.[5] Relative to market economies, there is also a far more limited range of choice of available housing types, styles, and quality levels. Moreover, the spatial distribution of housing in Eastern European cities often exhibits a pattern heavily influenced by huge prefabricated housing estates in the outer zones of the cities, frequently resulting in more dispersed housing than that found in market economies, and resulting in higher costs of commuting to and from work, higher infrastructure costs, and higher energy costs. The resulting disequilibria between household preferences and housing outcomes creates much of what appears to be high 'excess demand' for housing which

Table 15.9 Different measures of the housing situation around 1990 in the cities of three groups of countries

Indicator	Country grouping		
	Eastern Europe	Income comparators	Western Europe
Per capita GNP ($/year)	2,552	2,431	19,792
Share of owner-occupied housing (%)	28% (cities) 58% (countries)	62%	65%
Rent to income (%)	5.7	20.5	15.9
House price (median dwelling) to income	9.3	4.2	4.7
Floor area per person (m²)	19.6	14.0	32.3
Households per dwelling unit	1.12	1.16	1.02
Persons per room	1.28	1.74	0.67
Dwelling units per 1,000 people	366	207	481
Housing production: dwellings produced per 1,000 people	4.5	7.4	6.1
Housing investment as a share of GNP (%)	3.7%	6.3%	3.8%
Housing credit portfolio[a]	9.7%	15.6%	26.2%

Sources: Unweighted city data from around 1990, partly from the Extensive Housing Survey (World Bank), partly from the East Central European Regional Housing Indicators Database (MRI 1996).

Note
a   Defined as the ratio of the value of total housing loans to the value of all outstanding loans in both commercial and government financial institutions.

is reflected in high sales prices of housing relative to typical incomes, large black market premia for suitably located rental housing, and long waiting lists for state subsidized housing. These disequilibria impose costs not only on those whose preferences cannot be satisfied by the distributional system of the Eastern European housing model, but spill over into other areas of the economy as well, affecting negatively, in particular, labor markets – increasing regional wage differentials, distorting incentives for employers, and increasing levels of unemployment.

(Hegedüs *et al.* 1996: 106)

To sum up we suggest the following hypothesis: housing will not be among the major obstacles for the extension of the EU, because an acceptable minimum level of housing provision already exists in the countries of the East-Central European region. The serious problems with the distribution and quality of housing suggest, however, that there will be good opportunities for western institutions to invest in the improvement of the housing and urban sector of the region, especially taking into account the high expectations of population groups to live in better houses.

Housing problems and policies in the EU and their relevance for the Eastern European countries

EU integration and the regulation of housing

The main principle regarding the Community's approach to housing is as follows:

> In areas which do not fall within its exclusive competence, the Community shall take actions, in accordance with the principle of subsidiarity, only if and in so far the objectives of the proposed action cannot be sufficiently achieved by the Member States and can therefore, by reason of the scale or effects of the proposed action, be better achieved by the Community.
>
> (Maastricht Treaty 1992, see McCrone and Stephens 1995: 181)

The necessity for co-ordination, or problems of scarce resources or disproportionate burdens would justify Community actions, but housing does not belong to these categories.

McCrone and Stephens (1995: 188) conclude:

> it would be mistaken to accede to pressure to give the EU competence in housing policy or to think in terms of European funding directly for housing. There are no good grounds for supposing that policy can be more effectively operated at the level of the Union than that of Member States, or that the Union can achieve things that the states individually cannot achieve. There seems a clear case, therefore, for applying the principle of subsidiarity.

However, this does not mean that there is nothing to analyse. EU policies, the Structural Funds and European economic integration all have an effect on housing. The detailed analysis of the indirect effects of these programmes, however, is beyond the scope of this chapter.

Some critical housing issues in the unified Europe

In this section we raise some of the housing-related issues currently debated in the EU which may also have important consequences for the countries who want to join. At this stage we can only list these problems, rather than give a deep analysis.

Single market in mortgage lending?

McCrone and Stephens (1995: 218) show that 'inefficient' separated markets can be cheaper than a single market and that economies of scale are maximized below national level.

292

This is an especially important problem for the East-Central European countries, where housing finance systems are still very underdeveloped. An efficient mortgage lending system is missing and there are also huge problems with construction period financing (for a detailed description of the problems see Struyk 1996: 34). Thus there is potentially a big market for the western financial institutions. However, foreign institutions face serious obstacles to an easy eastward extension of their activities: the solvent demand is on average low and substantial buying capacity is restricted to the thin upper layers of society, and there are still legal problems with the practical application of enforcing methods, such as eviction and foreclosure.

Housing and the international and inter-regional migration of the population

Free labour mobility is a key assumption of a single market and this is clearly connected with the flexibility of the main tenure forms. This connection, however, is not simple, as empirical evidence shows substantial inflexibility of unskilled labour and also the existence of local/regional sub-markets (for example in housing) within countries. The most discussed issue in this regard is the link between immigration and housing policy (McCrone and Stephens 1995: 232). There are different patterns among the EU countries: in France and Germany immigrants exert substantial pressure on the social rental sector, while in Austria they do not, because they simply do not get title at all on social rental housing.

The flexibility of tenure forms in the East European countries is very different. In some of the CEE countries there is already now a lively real estate market for owner-occupied housing while in other countries the problems of title-registration, availability of mortgage, lack of real estate information, etc., are serious impediments to a better functioning real estate market. In some of the Eastern European countries the real estate markets are still closed to foreigners (as a form of protection against rich buyers pushing up real estate prices to a level domestic buyers cannot afford). There are also arguments against the free mobility of labour because the relatively better-off countries of the region intend to protect their internal labour market from the large masses of unemployed population in the neighbouring eastern countries.

The social rental sector in the East European countries shows very substantial inflexibility as not only are immigrants excluded from the local allocation of public rental housing but also regional migrant families (they usually have to prove several years' work or residence in the given settlement before getting the right to apply for public rental housing).

The necessity of regional planning

McCrone and Stephens (1995: 227) show the huge role regional policy plays in promoting flexibility in the housing market. Regional development policy is important, because:

> those who take the investment decisions that determine the geographical pattern of economic activity do not themselves directly bear all the social and private costs they entail. . . . Regional policies encourage new investment and growth to take place in areas where surplus resources are available, especially labour.

As a consequence of the political push for rapid decentralisation and privatisation, and also due to the political will to eliminate the previously dominating central planning institutions and systems, urban structures became very imbalanced in most of the East European countries. Not only did politically determined long-term planning disappear almost totally but so did the very useful and necessary medium-term planning. Most local authorities prepare only yearly budgets without any further outlook on the consequences of their decisions or on the actions of other actors.

Public expenditures and housing policy

The well-known inflation, unemployment and budget deficit data clearly show the bad public finance situation of those East-Central European countries where the transition towards a market economy already started. The budget deficit in percentage of GDP is very different from country to country and, interestingly enough, in some of the eastern countries this indicator is smaller than in many of the EU countries: Bulgaria −11.5 per cent, Hungary −7.5 per cent, Poland −2.8 per cent, Slovakia −5.7 per cent, Czech Republic +1.0 per cent, Romania −4.3 per cent (source: ECE 1995: 171). The same applies for the unemployment figures: Poland 16 per cent, Slovakia 14.3 per cent, Hungary 10.4 per cent, Czech Republic 3.2 per cent (ECE 1995: 111).

There is a growing recognition in the EU countries that not only should the targeting of subsidies be examined when analysing the social effects of public policy, but also the targeting of tax exemptions (McCrone and Stephens 1995: 238). This topic is currently not yet discussed in most of the East-Central European countries as the introduction of tax relief has only just begun to be talked about.

Another new trend in the EU comes from the idea that public housing should not rely solely on public sector financing. As a result, there are growing efforts to push housing companies to borrow from the private capital market (McCrone and Stephens 1995: 239). The East-Central European countries are very far from this understanding: instead of building up an effective non-governmental public sector, currently they are working on the reduction of their existing (in most cases already small) public rental sector through give-away privatisation.

In both parts of Europe the home-ownership rate is increasing and, in fact, as a result of quick privatisation some East-Central European countries have higher ownership shares (Albania 98 per cent, Bulgaria 90 per cent, Romania 89 per cent, Slovenia 88 per cent, Hungary 86 per cent, Croatia 85 per cent, Lithuania 79 per cent) than most western countries. It is a very interesting topic for comparison, which aspects of the high home-ownership rates can have distorting effects on the economy of average households. In Western Europe the role of mortgage loans is much more substantial than in East-Central Europe, therefore countries with high ownership rates (e.g. Britain) are very sensitive on economy-regulating measures. In the eastern part of Europe it is not the extensive spread of mortgages that is the problem but just the opposite: having virtually no access to mortgages (because of low incomes and high inflation) many families who very recently became home-owners are getting into very difficult situations in order to try to cope with the growing problems (e.g. deferred maintenance, increasing utility and renovation costs) in relation to freshly privatised multi-family buildings.

In fact, there are substantial differences between the western countries and the same is true for the East-Central European countries. For example the particularities of the British housing market and policy (dominated by owner-occupation) lead to destabilising effects on the economy. This 'volatility effect' is much less substantial in the case of Germany or France, due to the lower proportion of owner-occupation, lower turnover rate, lower level of personal debt and lower level of variable interest rate mortgages. Some of the East-Central European countries (notably Slovenia, Hungary) are now moving exactly in the British direction: owner-occupation has become very high, the turnover rate is increasing, personal debts are starting to increase and variable interest rate mortgages have been introduced. In some of the other countries (e.g. Czech Republic, Slovakia) market-oriented changes are much slower to gain ground, leading probably to a less volatile model (which, however, also has its serious problems mainly in the inefficiency of the unchanged institutional structure).

Summary: housing aspects of the eastward extension of the EU

The aim of this chapter was to address three questions regarding the extension of the European Union into Eastern Europe. Concentrating on the housing policy point of view and based on our database we can reach the following conclusions.

EU considerations: why extend to the east?

The decision about the EU extension is, of course, not a housing matter. Among the main arguments for the extension we can mention the desire to push the immediate border (the limit of the safe area) of the EU to the east and the belief that the extension will create big new markets for many types of services.

As a consequence of the fact that housing is not a strong integrational factor, the extension of the EU has no particular dangers from this point of view. Those countries which are the most probable first candidates do not have extraordinary housing problems. The main housing problem facing these countries is not absolute housing shortage, but more qualitative types of problem (such as the low efficiency of the housing finance systems, the huge deferred maintenance in the urban multi-family stock, the low level of infrastructure outside the big cities) and affordability problems. Thus if the extension takes place, no big waves of migration are to be expected (at least not for housing reasons) but a big demand for more efficient housing finance systems, for urban renewal programmes and for means-tested targeted housing benefits. These are, of course, mainly the tasks of the nation-states and will only affect the EU in so far as these problems concentrate on some less developed regions. There will obviously be a demand for the EU programmes subsidising the underdeveloped regions – however, the fears about the collapse of the EU budget because of extension to the east seem to be exaggerated, and this can be avoided with careful analysis regarding the Structural Funds in the new situation. On the other hand, western housing companies and financial institutions will get good investment possibilities in the region:

- for financial institutions to develop mortgage instruments and long-term saving options for renovation and special methods for financing new construction in high inflationary environments
- for brokers to modernise the real estate industry to service the growing number of real estate transactions
- for the building industry to introduce energy-saving measures (especially in the case of the huge prefabricated housing stock with outdated insulation and district heating systems).

All these measures must, of course, take into account the low solvent demand of the population. The example of the quick eastward extension of the German *Bausparkasse* (contract-saving) system, however, proves that this market is worthwhile to deal with for the developed western institutions.

East European countries: why join the EU?

For the Central and Eastern European countries there are clear political and economic reasons to join the EU. It is the immense interest of most of the new democracies to join a bigger Community where they can feel safe from unwanted tendencies (e.g. ethnic wars in their neighborhood) and can become part of bigger economic co-operation (custom-union, free movement of labour, etc.). As already mentioned, to put these countries into such a safe position and ensure the circumstances for their economic development is also in the interest of the EU.

Even if housing policy is not unified within the EU, and there are very different solutions to the housing problems across the member states, there is a continuous information exchange among these countries. New candidates for EU membership feel the challenge to modernise public housing and urban planning policies and they need 'technical assistance' in this regard. Existing tendencies in Eastern Europe towards extreme decentralisation and give-away privatisation can easily lead to a very difficult situation in the housing sector from which the way out must be based on a careful re-evaluation of the role of the public sector. There are important lessons to learn from the EU experience, e.g. regarding the necessity to keep a substantial and well-organised, efficient non-profit housing sector. Another area for learning from the western practice could be the experience with different types of subsidies (e.g. tax reliefs) and taxation methods (e.g. real estate tax). Many of the East European countries are about to introduce such measures, which have a long history in Western Europe with considerable knowledge about their advantages and disadvantages.

There are, of course, also some fears on the side of the institutions and the population of the Eastern European countries about the possible negative effects of integration into the EU. The emerging financial institutions of the East-Central European countries are afraid of the competition the much more developed western financial institutions would cause. There are also discussions among population groups and real estate experts about the potential consequences of opening up the real estate market to foreign buyers: there is a chance of a mass influx of rich western buyers pushing up real estate prices. This is a real dilemma, and not even the solutions applied in the case of the new member states of the EU (e.g. Sweden, Austria) can serve as examples, because these countries have relatively high real estate prices preventing the influx of average income foreign buyers.

Which countries should join the EU first?

The short answer could be: those countries which are the best prepared to join, whose populations want this step and which are also accepted by the existing group of countries of the EU. Housing will obviously not be a decisive factor when measuring the 'preparedness' of a country to join. Our database gives a good background to show the most important differences in the housing situation between the different groups of countries who are candidates to join the EU. On the basis of the existing data and from an overview of policy development we can raise the hypothesis that the CEE countries are more prepared for integration than the NEE or the SEE countries. In the CEE countries not only have the first steps of a market-oriented change of housing policy (privatisation of public stock, withdrawal of budget subsidies) been carried out, but also some efforts to introduce new regulations and institutions to address the new challenges (e.g. condominium law, mortgage regulations and institutions, inflation-adjusted loan products, computerised real estate and land registration systems). However, even the CEE countries have a long way to go to address their growing affordability and housing renewal problems.

Another important factor is the kind of differences among the Central and East European countries regarding the desirability of joining the EU. The available information (based on public opinion surveys) shows that the desire in the population to join the EU is the highest in the CEE and SEE countries, while data on the acceptance of the countries by the population of the EU show preferences for the CEE and Baltic countries.

From our analysis it can be seen that there are some open questions on both sides regarding the housing-related consequences of the eastward extension of the EU. Current member states of the EU have a fear that the big demand for regional and structural funds will lead to the collapse of the EU budget, and that large masses of the population will look for jobs in the west. Banks of the East-Central European countries are afraid of the competition the much more developed western financial institutions would cause, and the population is afraid that rich western buyers will buy up the best parts of real estate in the Eastern European countries, and so on. It is certain that these considerations will be part of the discussions on the joining of the new countries to the EU.

The main conclusion of our analysis could be, however, that besides some problems, there are much bigger opportunities on both sides in connection with the eastward extension of the EU. The housing markets of the new member states will substantially increase the investment possibilities of the western financial institutions, housing and construction companies. And hopefully the local and central authorities of the East-Central European countries will get more advice and technical help to solve their housing problems, just as the present member states exchange information and good practice, despite the fact that housing is not regulated at the level of the EU.

Notes

1 Many ideas in this chapter are based on joint research with József Hegedüs.
2 The 1994 data do not yet show the increased utility prices; since then, in most countries, there have been further sharp utility price increases.
3 Countries included, in increasing order of 1990 GNP per capita: Spain, United Kingdom, the Netherlands, Austria, France, Germany, Sweden, Norway and Finland.
4 Comparator countries include, in increasing order of 1990 GNP per capita: Jordan, Colombia, Thailand, Tunisia, Jamaica, Turkey, Chile, Algeria, Malaysia, Mexico, South Africa, Venezuela, Brazil, Korea and Greece.
5 There are neither commonly accepted measures nor reliable data on the problem of 'deferred maintenance'. To highlight the magnitude of the problem in one concrete example: in one of the inner city districts of Budapest there are buildings built 80–100 years ago and never really renovated since then. The current market value

per square metre of the flats in the run-down buildings is around US$220–250. The renovation of the building, i.e. the solving of the deferred maintenance problem (without renovating the flats inside) would cost approximately US$220 per square metre. As a result, the non-renovated flats in the renovated buildings would have a market value of US$360–400. It can be seen that the costs of renovation necessary to handle the problem of deferred maintenance are higher than the value-increase of the flats. Additionally, it is clear that tenants currently occupying these flats cannot be forced to pay rents which would cover renovation costs.

References

ECE (1995) *Economic Survey of Europe in 1994–1995.* New York and Geneva: United Nations, Economic Commission for Europe.

EU (1993) *Statistische Daten über das Wohnen in der Europaischen Gemeinschaft.* Brussels: EU DG V.

Hegedüs, J., Mayo, S. and Tosics, I. (1996) 'Transition of the Housing Sector in the East Central European Countries', in *Review of Urban and Regional Development Studies* 8: 101–136.

Hegedüs, J. and Tosics, I. (1996) 'The Disintegration of the East-European Housing Model' in D. Clapham, J. Hegedüs, K. Kintrea and I. Tosics (eds) *Housing Privatization in Eastern Europe.* Westpont, Conn.: Greenwood Press.

McCrone, G. and Stephens, M. (1995) *Housing Policy in Britain and Europe.* London: UCL Press.

MRI (1996) East Central European Regional Housing Indicators Database. Metropolitan Research Institute, Budapest.

Pickel, G. and Pickel, S. (1995) 'Attitudes towards European Integration and Democratic Stability in Eastern Europe in Comparison', paper prepared for the second European conference for sociology: 'European Societies: Fusion or Fission?', Budapest.

Struyk, R. (1996) 'The Long Road to the Market', in R. Struyk (ed.) *Economic Restructuring of the Former Soviet Bloc. The Case of Housing.* Washington, DC: Urban Institute Press.

UN (1995) *Trends in Europe and North America. The Statistical Yearbook of the Economic Commission for Europe.* New York and Geneva: UN.

Wallace, C., Chmuliar, O. and Sidorenko, E. (1995) 'die östliche Grenze Westeuropeas. Mobilitat in der Pufferzone', *SWS-Rundschau* vol. 35, 1: 41–69.

INDEX